CAMBRIDGE LIBRARY COLLECTION

Books of enduring scholarly value

Life Sciences

Until the nineteenth century, the various subjects now known as the life
sciences were regarded either as arcane studies which had little impact
on ordinary daily life, or as a genteel hobby for the leisured classes. The
increasing academic rigour and systematisation brought to the study of
botany, zoology and other disciplines, and their adoption in university
curricula, are reflected in the books reissued in this series.

The Amateur's Flower Garden

One of the most popular and successful gardening writers of the Victorian
era, Shirley Hibberd (1825–90) was editor of three bestselling gardening
magazines. He was highly influential – one of the first to highlight issues such
as environmental conservation, water recycling and cruelty to animals – and
he helped to establish what is now the vast consumer industry of amateur
gardening. First published in 1871, this is one of many books he wrote on
the subject, intended as a 'handy guide' for the creation of attractive flower
gardens. In it Hibberd offers advice on such topics as bedding plants, border
flowers, rockeries, and annual and biennial plants. He also presents methods
for managing various types of garden, such as subtropical, alpine and rose
gardens. Highly detailed and extensively illustrated, this book remains useful
and relevant to both amateur enthusiasts and seasoned horticulturists.

The Amateur's Flower Garden

*A Handy Guide to
the Formation and Management
of the Flower Garden and the
Cultivation of Garden Flowers*

SHIRLEY HIBBERD

CAMBRIDGE
UNIVERSITY PRESS

CAMBRIDGE UNIVERSITY PRESS

Cambridge, New York, Melbourne, Madrid, Cape Town,
Singapore, São Paolo, Delhi, Mexico City

Published in the United States of America by Cambridge University Press, New York

www.cambridge.org
Information on this title: www.cambridge.org/9781108055345

© in this compilation Cambridge University Press 2013

This edition first published 1871
This digitally printed version 2013

ISBN 978-1-108-05534-5 Paperback

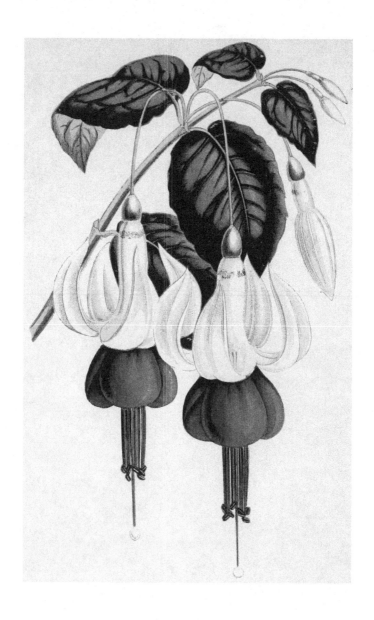

SHOW FUCHSIA.

(White and rose.)

THE

AMATEUR'S FLOWER GARDEN.

A HANDY GUIDE TO

THE FORMATION AND MANAGEMENT OF THE FLOWER GARDEN.

AND

THE CULTIVATION OF GARDEN FLOWERS.

BY

SHIRLEY HIBBERD,

AUTHOR OF "RUSTIC ADORNMENTS FOR HOMES OF TASTE," "BRAMBLES AND BAY LEAVES," ETC.

Illustrated with Coloured Plates and Wood Engravings.

LONDON:

GROOMBRIDGE AND SONS,
5, PATERNOSTER ROW.
MDCCCLXXI.

LONDON :

PRINTED BY SIMMONS & BOTTEN,
Shoe Lane, E.C.

CONTENTS.

THE

AMATEUR'S FLOWER GARDEN.

INTRODUCTION.

> " Maud has a garden of roses
> And lilies fair on a lawn ;
> There she walks in her state
> And tends upon bed and bower.
> And thither I climbed at dawn,
> And stood by her garden gate ;
> A lion ramps at the top,
> He is clasp't by a passion flower."
>
> TENNYSON.

A FLOWER GARDEN is intended for the cultivation and display of flowers; but any book upon the subject, however small, must treat of other matters, not as subordinate to the leading idea, but as necessary accompanying features. Hence, in the chapters that follow, some attention is paid to the shrubbery, the lawn, the walks, the greenhouse, and the window; for to pass them by, in order to treat of flowers only, would be to court imperfection, while, to bestow over-much attention on them would be to thrust into a secondary place the very feature that should take the lead. It will be understood, therefore, that this book, though a very small one, is at least comprehensive in purport, and aims at providing its possessor with useful guidance in the formation and management of the flower garden, according to the generally accepted meaning of that term. It might have been entitled, "The Pleasure Garden in Little," but its object and scope will, no doubt, be better understood by the simple and commonplace title that has been adopted. As gardens vary in extent, in charac-

1

teristic features, and in the requirements of privacy in one place and public display in another, it is simply impossible, in a work of such limited dimensions and pretensions as the present, to attempt an exhaustive treatment, either of the whole subject, or any one of its more important constituent parts. It must be understood, then, that while the attempt is made to gratify a variety of tastes, and accommodate a number of different circumstances, a somewhat contracted boundary of the field of operations is kept in view from first to last. In other words, if this book should prove useful at all, it will be to such as possess what may be called "homely" gardens as distinguished from great and grand gardens, and especially from gardens that are kept for purposes of show.

It is, above all things, necessary in a book of this kind, to recognize at every step the requirements of nature, and the best established principles of art as distinct altogether from individual taste and fancy. If it is herein stated that roses will not grow like house-leeks on tiled roofs, nor rhododendrons in beds of chalk, those points must be considered settled, for they do not admit of discussion. But when it is further added that beds of roses do not assort tastefully with beds of geraniums, that coniferous trees are out of place in a flower border, there is room for difference of opinion, and the reader is at liberty to quarrel with the author to any extent, and set at nought every one of his advices and suggestions. Perhaps there will be less said about taste than practice in the following pages; but it is a difficult matter to write on a subject which has occupied one's attention, both as a business and a hobby, for a quarter of a century, and on many matters connected with which distinct opinions have been formed, without being occasionally betrayed into expression of those opinions, or, at the least, of indicating the direction in which intentionally-concealed opinions tend. On matters of practice, the practical man has within certain limits which propriety will point out, the right to dictate. On matters of taste, dictation is equally unjust and absurd. When we encounter subjects that divide opinions amongst those who study them, we must be careful to avoid dogmatism, and that spirit of self-satisfaction which would make "I say" a law binding on all the world. But when the range of opinion is limited, and its limits are appreciable only by the aid of technical knowledge, it is another matter, and the man who knows may proceed to lay down directions for those who need them, provided he will always keep in mind to be correct, and explicit, and as brief and modest as possible. On these principles I propose to labour in the preparation of this and other works intended to follow it; and I make the declaration at starting that, as regards principles, there shall never, through any carelessness on my part, arise the shadow of a misunderstanding between me and my readers. I shall have to deal chiefly with matters of fact, and hope always to have the discernment to keep them distinct from matters of fancy.

S. H.

GARDEN-SEAT BY DEANE AND CO.

CHAPTER I.

FORMING THE FLOWER GARDEN.

WHATEVER the dimensions, the position, and the purpose of a flower garden, whether for private enjoyment or public display, perfect success in its formation and management cannot be insured, unless a few necessary conditions are complied with. We may find examples in abundance of good and bad gardens, and shall not be long in making the discovery that a great display of flowers is not alone sufficient to afford the pleasure which a cultivated taste will always expect as the proper reward for the expense and care that have been incurred in its production. During the past twenty years there has been a constantly-increasing tendency to superficial glare and glitter in garden embellishment, to the neglect of the more solid features that make a garden interesting and attractive, not only to-day and to-morrow, but " all the year round." The magnificent displays of bedding plants in our public parks and gardens have, without question, favoured a false estimate of the proper uses of gardens in general. We have seen the development of an idea which, in consequence, regards private gardens as exhibition grounds, and tender plants of the geranium, verbena, and petunia type as their only proper occupants. Now, it will be our business in subsequent chapters to treat upon the bedding system, and the plants that constitute its primary material elements ; but it is important here, with the question of forming the flower garden before us, to take note of the fact that the modern flower garden, as known to tens of thousands of persons, is a poor, gingerbread-entity, ephemeral in respect of its best features, and while demanding but little talent for its production, offering an equally small return in the way of intellectual enjoyment. Before flowers are thought of, a garden should be provided for the sustenance of a suitable extent of shrubbery, grass-

turf, and other permanent features, to which the flowers will
in due time serve for embellishment, and, in return for this
service, have the advantage of a sufficient extent of leafage
and verdure to heighten their beauties by harmonious surroundings. A garden rich in trees and shrubs, with ample
breadth of well-kept lawn, will be enjoyable at all seasons
without the aid of flowers. A few simple borders, well
stocked with mixed herbaceous plants, such as primulas,
pæonies, lilies, phloxes, hollyhocks, and carnations, would, in
many instances, afford more real pleasure and ever-changing
interest than the most gorgeous display of bedding plants
hemmed in between two glaring walls, or exposed on a great
treeless, turfless place like the blazing fire at the mouth of a
coal-pit. But given the good permanent substratum, the
well-kept garden of greenery, with its family trees and its
interesting plants that one can talk to, and its snug nooks
filled with violets and primroses, and its mossy banks that en-
tice the early sloping sunshine, and its cool coverts, where ease
may be enjoyed amid the summer's heat, and then a brave
display of flowers becomes the crowning feature. The argument may be summed up in this—that flowers alone do not
constitute a garden ; and when a garden has been provided
to receive them, the display should be adapted in extent
and character to the situation and its surroundings.

A considerable number of features are recognized as proper
to a flower garden. In respect of formation and management,
these may be considered as separate and distinct, and hereafter
it will be necessary to isolate them. But in the general plan
they should all be intimately related, as natural and necessary
developments of a comprehensive idea. The outer boundaries
of tree and shrub, the intersecting walks, the belts of evergreens, the mixed borders, the air-inviting lawns—these combine in their relationships to create the want of a parterre ;
and if the garden is one of ample extent, several distinct
displays of flowers, or rather several little gardens, will be
admissible in consistency, and may be desirable for the
occupation and entertainment of the owner.

At this point it seems needful to unfold some elaborate
plans, but it will be safer to say that the compass of the
book does not admit of them, and that they would be more
proper to a treatise on the " Pleasure Garden," which this is
not; for it is only one department of the pleasure garden

that really concerns us. A few plans may, however, be useful
here, as affording suggestions; and we offer them with the
qualifying remark that every separate garden needs a separate
plan adapted to its dimensions and position, and therefore
ready-made plans are but of secondary value. The two grand
requirements of the design for a garden necessitate a special
consideration of every special case. And what are those
two chief requirements? To my thinking, the plan of a,
garden should be such as to develop to the utmost the capabilities of the site, and represent the particular taste and
fancy of the owner. Whatever is attempted should be within
the possibility of a successful result, and no one should make
difficulties without first counting the cost. At every step the
wise gardener will ask Nature what she thinks about it. The
result will be equal avoidance of mistakes and attainment of
successes. Standard roses planted on grass turf, without any
space of open soil around them, *never* thrive. Yet everywhere we see examples of this ridiculous blunder, and
entrance-courts that might be rich and stately are made
hideous with the starving sticks ostentatiously stuck about
the turf. Rhododendrons will not thrive in clay or heavy
loam, yet everywhere we see them planted with laurels,
aucubas, and such things, to last only as long as the ball of
peat planted with them suffices for their support, after which
they shrivel up, and, unless removed and burnt, disgrace the
garden. Bedding plants, almost without exception, require
to be fully exposed to sunshine, yet we see them planted in
shady places, where they soon become sickly, and cease to
flower, though those very same shady spots might have been
made beautiful by means of flowers that need not full exposure to sunshine. Every garden design, and every project
of garden furnishing, and every item of garden work, should
be governed by the consideration that it is hard work to fight
against Nature, and there is never a prospect of a conquest
worth obtaining. Those who will aim at development of the
capabilities of a garden will, in spite of the mistakes and
misfortunes that attend all enterprises, be pretty sure to
secure enjoyment in the end. Fortunately, if gardening is
pursued with earnestness, every soil, every situation, the
breezy hill-side and the smoky city, will be found to have
some capabilities which art can turn to account by patiently
accepting the teachings of Nature.

In laying out a garden, it is impossible to foresee what changes it may undergo as new wants arise, or as fancy, seeking a homely field of exercise, may dictate as " improvements." It is therefore well (except in particular cases that need not be provided for) to adopt in the first instance a simple plan that will afford a fair basis for after elaboration, as circumstances arise to necessitate it. Such a plan, in skeleton, is here figured. It was drawn for a friend who had taken a piece of rough, low-lying meadow-land, on which to build a house and make a garden. It is drawn on a scale of fifty feet to the inch. D R is the drawing-room, the windows of which look upon a small neat lawn, dotted with coniferous trees and clumps of rhododendrons. T is the terrace; F T, plantation of fruit-trees; K G, kitchen-garden plots. The dotted lines show the course of the drain-pipes, the land falling away from the house somewhat rapidly. The conservatory, c, and the boundary borders, s, need not be remarked upon, but the other features demand a few words. In the first instance, the ground presented a steeper slope than was desirable, and being a clay soil heavily charged with moisture, the highest part was selected for the house, and that was raised considerably by means of the earth taken out for the foundation. Thus was formed the terrace, an excellent feature, for it commands an extensive view over a beautiful piece of country, which was scarcely visible from the same spot, until a higher level was obtained for the advantage of the house. The outlying E H is an engine-house, which is quite excluded from the terrace view by means of a few trees planted for the purpose. The lawn is, of course, on a dead level, but beyond that point the ground falls gently to the boundary in the rear, where there is ample outlet for the drainage. Let us suppose now that the proprietor takes in another piece of land for fruit and vegetable culture, or gives up those things for the sake of flowers. The plots below are available for any scheme consistent with the capabilities of the place. On the pieces marked F T may be formed a geometric garden, enclosed by clipped hedges of yew, arbor vitæ, or by a fence covered with climbing roses. On the K G pieces may be formed a mixed flower garden for hardy herbaceous plants, roses, and flowering shrubs. And the extreme rear plot marked B B, for bush fruits, may be planted with a mixture of the most elegant low-growing, deciduous trees, to make a fringe of wood to

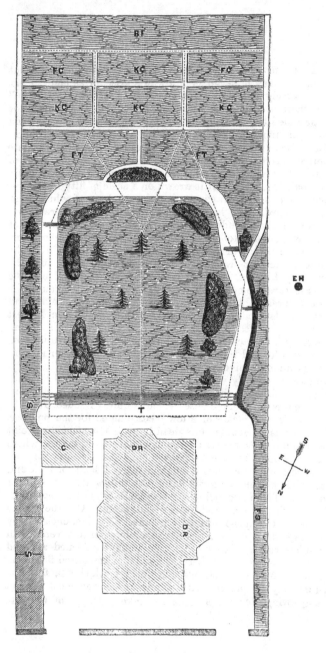

SKELETON PLAN FOR A VILLA GARDEN

mark the extent of the property without obscuring the view
over the country from the terrace.

The next example is a complete plan, adapted to a peculiar
conformation of ground.   It represents a beautiful and interesting garden, the completeness of which has been arrived at

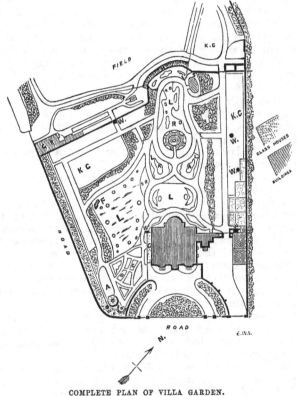

COMPLETE PLAN OF VILLA GARDEN.

by means of successive alterations and expansions of the
original skeleton plan.   A few remarks on this will, no doubt,
be acceptable to the reader.

In the formation of a garden plan, one of the chief
requisites, a good supply of water, must be considered—and

within reasonable limits the more watering-places the better. Such are marked (w) in the accompanying plan. With the aid of connectable lengths of gutta percha pipe they are found sufficient.

The front garden being only separated from the high road by light iron railings, is principally stocked with evergreens, the border being filled with bedding plants. A screen of trees effectually divides the vegetable garden (K G) from the flower parterres and lawns (L); G R the gardener's residence; H, a hawhaw, separating field and plantation from flower garden ; R, in the centre of the plan, is a rockery, encircling a basin containing gold and silver fish, a raised central vase, and fourteen grouped fountain jets. A summer-house (S) is situate at the bottom of the garden next the field. From hence a view of the open country is obtainable. S R is a bed devoted to standard roses. D G shows a geometric garden situate near the greenhouse ; F is a fernery with rockwork arches. The circles on the lawns denote positions of favourite trees ; O is an octagonal greenhouse, in Crystal Palace style. In the centre are tables and chairs, and the gas being laid on, it is a favourite resort for summer evenings.

At the side of this conservatory is another fountain and fernery, the former being supplied from a tank hidden among the trees. A waste pipe in the upper vase, forms the means of supply to a small overshot water-wheel in the rockwork. From this wheel the water flows to a lower basin. G is a greenhouse in connection with the residence—at one end is a collection of exotic ferns, ten jet fountains, miniature cascade and turbines—at the other end is an ironwork fountain, with ornamental basin.

The forcing and orchid-houses are to the extreme right, and hidden by a screen of trees and shrubs. A shows the position of American beds stocked with hardy rhododendrons, azaleas, kalmias, andromedas, and heaths.

The limits of this work will not admit of any elaborate disquisition on the principles of taste in gardening, or on the mechanism of garden construction; but a few practical suggestions may prove useful to many readers who desire to form new gardens or improve old ones. It is desirable in the first instance to secure good roads and walks, good lawns, and good shrubberies, before thinking much about flowers. These three primary elements should be provided in the best form

possible, and with such a forecast of possible future operations
that none of the work shall have to be undone during the
remainder of a lifetime. The system of drainage should be
ample, and all the measures adopted to remove surplus water
from the ground should have their counterparts in measures
devised for putting water on when required. Thousands of
people can show us bright flowers in summer time, in juxta-
position with grass turf burnt to the semblance of a worn-
out mat. Generally speaking, bedding plants require no
water after they have had good nursing for a fortnight
after being planted, and the time usually wasted in keeping
them watered might be better employed in flooding the grass
periodically during droughty weather, with the aid of flexible
hose, connected with a supply adapted to the purpose. Very
much is thought of a south aspect, but for the enjoyment of a
garden from the windows, a north aspect is invaluable. You
look out during the whole of the forenoon on the sun-lighted
garden, from a cool, shady room, and nine-tenths of all the
flowers that occupy the view turn their faces towards the
window; or, to speak more correctly, look southwards, and
that practically is the same thing. Nor is it a small matter
to have a shady piece of turf in immediate contiguity with
the house, for conversation with friends, and for the games
that are proper to the summer season. In arranging a garden
with a view to the fullest development of its capabilities, it
is well to remember that, as a rule, evergreen shrubs will
thrive in partial shade, and a few of them in profound shade;
that flowers, as a rule, need the fullest exposure to sunshine,
though the exceptions to this rule are many, while grass turf
will thrive in sun and shade, if nowhere heavily shaded, and
may be employed to connect and harmonize all kinds of scenes,
from the highly artistic to the extremely rustic. It is desirable that every garden should present a few distinct
features, or at least one feature, to give it a character of its
own. The owner must determine this matter by a consideration of the possibilities of the situation, the nature of the
means at command, and the particular taste to be gratified.

In respect of garden furniture, we can only find room in
this chapter for reference to Edgings, and this subject we
cannot pass, for trivial as it may seem, it is a source of much
trouble and vexation. In open breezy places, dwarf box
makes the best edging in the world, and the cheapest in the
end, no matter what its cost in the first instance. If a green

edging is desired in a spot shaded by walls or trees, box is
useless, but common evergreen euonymus will take its place
tolerably well.   Grass verges are beautiful, if well kept; but
they entail a lot of labour to keep them trim, and it is always
a question if the time spent in clip—clip—clipping them might

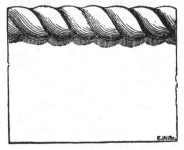 

not be devoted to something better.   Well-made edgings of
ivy have a solid, rich appearance ; but it would render a large
garden heavy in character, and an example of a good idea
overdone, to employ ivy edging everywhere.   There are three
sorts of substantial edgings available for different parts of
the garden.   If we consider the entrance-court first, we must
have either clipped box, clipped yew, or bold sharp bands of
ivy, or a handsome stone moulding, or its equivalent in some

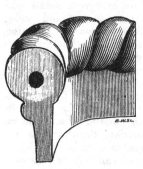

imitative material, such as Ransome, Rosher, or Austin can
supply.   These manufacturers
turn out substantial edgings of
artificial stone in an almost
endless variety of patterns, from
the extremely simple to the
most elaborate, those of an ornamental character being admirably designed.   An immense
variety of edgings are manufactured of tile, brick, and even
glass, and these are more or less
good, according to material,
manufacture, and price.   They are, as a rule, objectionably
frail ; they do not make sufficient foothold to keep true in
line for any length of time, and they are apt to crumble to
powder if a hard frost catches them immediately after heavy

rain or snow. The best cheap tile I have yet seen is one made for me by Mr. Looker, of Kingston-on-Thames, for supporting a border, which stands above the walk in an out-of-the-way part of the garden. It is in form an unequal triangle, nine inches high, and six inches broad, carefully made, and well baked. If set on a true, firm bed, it is practically immovable, and proof against all weathers. The border it supports is planted along the front line with a number of half-trailing plants, which hang over the sloping front, and form a varied fringe of beautiful vegetation, quite hiding the low red wall of tiles, which gives the border its definite boundary. When costly edging kerbs are put down, it is advantageous to bed them on brick footings, the top line of which should be an inch or so below the level of the gravel. This adds to their strength and immobility.

In planting choice shrubs amongst trees, it is advisable to take precautions against that warfare of competing roots by which frequently the undergrowth of a plantation is killed out. In the use of shrubs worth special defence, the plan shown in the figure is admirable. Dig a hole of a suitable size, say to measure a yard and a half deep, and a yard wide right and left; case the hole with brickwork, and at a third of the depth from the bottom let in a platform of stone or elm planks. On the platform lay down a bed of 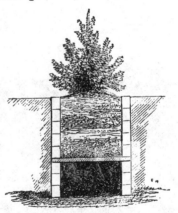 broken pots, then fill up with suitable soil, and plant the shrub. The vacant space beneath the platform will prevent the roots of the big trees working up into the good soil provided for the shrub. This is a costly mode of procedure, but in a spot required to be richly furnished, it is to be recommended, because it insures, amidst large trees, a free undergrowth of the most beautiful evergreen shrubs, provided only there is light enough to keep them healthy.

It is usually required of a writer on gardening to point out

how the principal features of a garden may be made to appear
greater and grander than they really are.   I feel bound to say
that while I would insure for every reality a due degree of
importance, I would, except in a few peculiar cases, oppose
the introduction of deceptions of every kind.   But it may
happen that a fantastic screen to hide an ugly object may afford
amusement to justify its adoption, and a humorous conceit in
a garden need not be of necessity despicable.   As an example,
therefore, of a pardonable trick, here is a figure of a screen
which bears the designation " elephant trap," in a part of a
garden which overlooks a road that no one in the house desires
to see.   The trees in the scene are real; but the contrivance
is a delusion—the screen being flat, and the seemingly long
winding path being taken up a gentle rise by a curve which
lengthens it without seeming to do so.   It answers its purpose,
and that is one proof of merit.

AN ELEPHANT TRAP.

CHAPTER II.

GEOMETRIC gardens may be designed on paper by selecting some part of the pattern of a carpet or wall paper, or by placing a few bits of coloured paper in the debuscope, and then copying the multiple scheme so produced. Numbers of designs have been obtained in that way, and about one in a hundred have actually turned out worthy; the rest were not worth the paper they were drawn on, unless it might be to make burlesque of the bedding system. It is a most rare event for a really complicated plan to prove effective, however skilfully planted; and so I begin this chapter by advising the beginner to avoid the schemes which combine a great variety of figures, such as ovals, hearts, diamonds, horns of plenty, and true lovers' knots. Elaborate designs are, of course, not to be contemned, for we find them constituting important features in many great gardens, and employing the highest artistic talent in garden colouring. It is above all things necessary, in an elementary book of this sort, to guard beginners against making costly mistakes, and the formation of the parterre is a business requiring more than ordinary caution to guard against waste of time and money, and all the consequent vexation and disappointment. In what we may call " a quiet garden" of limited dimensions, a few large beds, far separated by well-kept turf will, in many cases, give far more satisfaction than a distinctive geometric scheme, and necessitate, perhaps, only a twentieth of the time and attention to keep them suitably gay, besides offering the peculiar advantage that each bed may be planted to produce an effect of its own without any special reference to the rest, so long as it is decidedly different. The common repetition of oblongs and circles which we meet with in public gardens, where long walks demand flowery dressings, is one of the most effective and satis-

factory, though always open to the accusation of an alliance
with commonplace and monotony. On the other hand, the
common repetition, on the margins of lawns in private gardens,
of circular beds containing standard roses, surrounded by
geraniums, verbenas, and other such stuff, is ineffective and
puerile. Gardens embellished in this way have no character
at all, they are mere confusions. Far better would it be to
concentrate the energies which the "pincushion" beds con-
sume to a poor purpose, on a neat and reasonably circumscribed parterre, which would constitute a feature and afford
considerable interest. To be sure, it is easy to plant pincushion
beds, because they are scarcely co-related, but a parterre demands talent, and that is not always available.

In a majority of instances, geometric gardens are laid out
on grass turf, and the green groundwork adds immensely to
the beauty of the flowers.

In elaborately furnished
gardens, a groundwork
of silver sand, with box
embroidery to define the
outlines and fill in the
angles, is employed in
an open space set apart
for the purpose, and the
scheme is enriched with
statuary, clipped yews,
laurels, cypresses, and
vases containing yuccas,
agaves, or masses of ge-
raniums. The working
out of a great design
in coloured earths and
flower-beds is the most complicated, and, generally speaking,
perhaps the least satisfactory, form of the parterre. It has this
advantage, that, during winter, it affords " something to look
at," but the corresponding disadvantage is that nobody wants to
see it. A favourite idea with artists in this line of business is to
draw out, on a gigantic scale, a group of rose, shamrock, and
thistle in coloured earths and box embroidery, and while the
thing is new it looks tolerably well; but the majority of
people do not keep themselves sufficiently under control when
tempted to indulge a smile as they admire it. Generally

speaking, the design vanishes in summer, that is to say, when
the beds are full of flowers, the coloured earths that mark out
the design are so completely extinguished that, even with a
key plan in one's hand, it would be hard to see where the
thistle begins and the shamrock leaves off, and where, amidst
the confusion, the rose ought to be. The principal materials
employed for the intersecting walks in these designs, are
pounded Derbyshire spar (white), pounded brick (red),
pounded slate (blue), pounded coal (black), sifted gravel grit
(yellowish grey).

In planting the parterre it is as easy to make mistakes as
in designing it, and the most frequent errors are the employment of primary colours in excessive quantity and strength,
and the neglect of neutral tints to soften it, and of brilliant
edgings to define it. The stereotyped repetition of scarlet
geraniums and yellow calceolarias is in the last degree vulgar
and tasteless,
and the com-
mon dispositions
of red, white,
and blue are
better adapted
to delight sava-
ges, than repre-
sent the artistic
status of a civi-
lized people.
The increasing
use of leaf co-
lours marks a
great advance
in taste, and
strange to say,
the most perfect
examples of par-
terre colouring
we have seen of
late years, have

DESIGN FOR GEOMETRIC GARDEN.

been accomplished by leaves solely, in scenes from which
flowers were utterly excluded. Leaf-colours, however, are
of immense importance in connection with flowers, as any
good example of parterre colouring will prove. They afford

2

material for boundary lines, for relief agents, and for marking
the rhythm of combinations. Every scheme that is to be
viewed as a whole, must be coloured as a whole, and with the
object of producing a complete and harmonious picture. Whatever the nature of the materials employed, certain principles
must be followed to insure a satisfactory result. The strong
colours must be spread pretty equally over the whole scheme
with neutral and intermediate tints to harmonize and combine
them. The colours containing most light, such as yellow,
white, and pink, should be placed in the outer parts of the
design, to draw it out to its full extent; and the heavier
colours, such as scarlet, crimson, and purple, should occupy
the more central portions of the scheme. The most difficult
of all colours to dispose of satisfactorily is pure yellow, and its
related tints of buff and orange. A bed of yellow calceolarias in the centre of a group will be pretty sure to spoil it,
no matter how skilfully in other respects it may be planted.
But a few of the most conspicuously placed of the beds in
the boundary of the pattern may be planted with calceolarias
to assist in defining the arrangement. Bright and sharp
edgings are eminently desirable, and it is a good point if
the edgings are the same throughout, forming clear fillets of
silvery or golden leafage, or some suitable flowering plant,
which carries plenty of light in its colour. Objection may
be taken to this rule, on the ground that beds containing
plants that nearly approximate in tone to that of the general
edging, will be spoiled if edged like the rest. But the objection is superficial. When we cannot bring out the masses by
means of the edgings, and it is desirable to have the boundary
lines alike all through, we must change our tactics, and bring
out the edgings. For example, we are to suppose three beds
filled with flowers. No. 1 contains scarlet geraniums, and
may be edged with a band of blue lobelia, and an outer defining line of silvery cerastium; No. 2 is filled with blue
ageratum, and edged with a band of Purple King verbena,
with a finishing line of cerastium. No. 3 consists of Mrs.
Pollock geranium and blue lobelia, plant and plant, with a
finishing band of lobelia, and a boundary line of cerastium.
Thus, in three extremely different cases, the final fillet is the
same without violation of harmony or detraction from the
pronounced character of the beds. It is a matter equally important and interesting, that a perfect hypothetical balance of

colours is neither a good practical balance nor agreeable to the educated eye. A square yard each of red, blue, and yellow, whether in grass or gravel, will not make a telling parterre. But a block of blue, between two blocks of red, and all three banded with a silvery grey line or a sufficient breadth of green grass, might constitute an agreeable, though humble feature of a garden. It is well, indeed, in every scheme to allow one colour with its related shades to predominate, and to employ the others as relief agents rather than as features. Lastly, strong contrasts should not be indulged in often; they are the antitheses of harmony, as you may discover by observation. Thus we shall find two geraniums like Thomas Moore and Feast of Roses, the first intense scarlet, the second intense rose pink, produce a most delightful harmony when planted side by side. And again, Bonfire geranium, a dazzling scarlet, may be planted by the side of Purple King verbena, with the certainty of a rich and perfect combination. This much, however, must suffice on the subject of colour for the present; but we shall have to return to it in connection with the plants required for the bedding system. If example is better than precept, the best part of this chapter is now to come, for examples are needed; and the few selected are well adapted to illustrate principles.

The subjoined figure, p. 20, represents a panel garden, drawn to scale. It lies immediately below the terrace, and is approached by a flight of steps. On either side is a strip of grass, twelve feet wide, on the same level as the flower beds, and beyond that the ground rises in a grass slope (or ramp) to the general level of the lawn above. Two examples of planting this garden for a summer display will be given, and the first shall be a harmony in red. No. 1, Stella geranium, or an equally rich and heavy crimson scarlet geranium; 2, 2, Blue Lobelia, and a golden-leafed geranium, such as Golden Banner; 3, 3, a dwarf scarlet geranium, such as Attraction or Thomas Moore; 4, 4, 4, 4, same as centre; 5, 5, 5, 5, solid planting of a good rose-pink geranium, like Christine, or Feast of Roses. Nos. 4 and 5 being in the same boundary, and, in fact, one and the same bed, the scarlet must occupy the half nearest the centre, and the pink the other half; 6, 6, 6, 6, Amaranthus melancholicus, edged with Centaurea ragusina; 7, 7, Coleus Verschaffelti, with outer band of yellow Calceolaria; 8, 8, same as 3, 3, and edged with blue Lobelia; 9, 9, a pale pink gera-

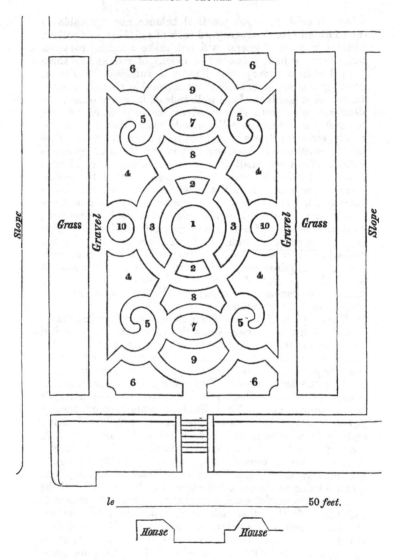

nium, such as Pink Muslin, or Rosa Queen; 10, 10, Geranium Avalanche, which has white leaves and white flowers. The second example of the planting shall be a harmony in blue. No. 1, Petunia Purple Bedder, or Spitfire, or Verbena Celestial Blue, edged with Cerastium; 2, 2, Dwarf Scarlet geranium, edged with blue Lobelia; 3, 3, a tricolor geranium, such as Sunset, or Louisa Smith, edged with blue Lobelia; 4 and 5, in centre of each division of these compartments, about where the figures are placed, a circular dot of a brilliant scarlet geranium, such as Thomas Moore, or Lion Heart, the rest of the block filled in with blue Lobelia, finished with edging of Cerastium; 6, 6, 6, 6, Geranium Flower of Spring, and blue Lobelia, plant and plant, edged with Ivy-leaved Geranium Elegant; 7, 7, a dwarf scarlet geranium, edged with blue Lobelia; 8, 8, Lobelia Indigo Blue, edged with Geranium Flower of Spring; 9, 9, a lilac or rose-pink geranium, such as Lilac Banner, Feast of Roses, or Amy Hogg; 10, 10, a dwarf salmon or orange-scarlet geranium, such as H. W. Longfellow, or Harkaway, edged with Cerastium.

The next example, p. 22, makes a poor appearance on paper, but in the fine large old-fashioned garden, where it embellishes the forefront of a lawn, it is a most effective arrangement, the beds being cut out on the grass, and all of them furnished to produce decisive effects. When the drawing was made, the beds were filled as follows: A, White Verbena, edged with Purple Verbena; B, Mangles' Variegated Geranium, edged with blue Lobelia; C, C, Lion Heart Geranium, edged with Flower of the Day; D, Crimson Unique Geranium, edged with Flower of Spring; E, E, Geranium Tristram Shandy; F, F, blue Lobelia, and Cerastium tomentosum, plant and plant, edged with Cerastium; G, Geranium Duchess; H, Geranium Louisa Smith; I, I, vases filled with Ivy-leaved geraniums, Gazanias, and Convolvulus Mauritanicus.

In further illustration of the principles of geometric colouring, a selection has been made of a series of schemes in the Liverpool Botanic Gardens, where Mr. Tyerman, the able curator, has developed this system of embellishment with peculiar completeness and success. The first of the series will indicate the value of geraniums, or, as they should be termed more correctly, zonate pelargoniums; for the whole furniture consisted, in the season when these notes were made, of varieties of this class of bedding plants, with the exception, as will be seen, of a few trivial dots of calceolaria and verbena.

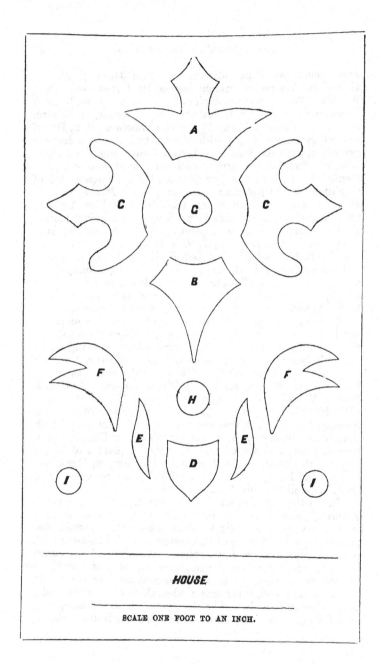

HOUSE

SCALE ONE FOOT TO AN INCH.

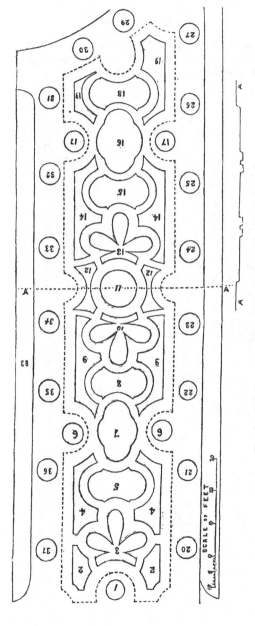

The sketch is drawn to scale, with a section of the levels through.

1. Purple trefoil, with centre of Chamæpuce diacantha, edged with Saxifraga longifolia. 2, 2. Geranium Queen of Queens. 3, 3. G. Christine. 4, 4. G. Little David. 5. G. Stella. 6, 6. Calceolaria canariensis and scarlet verbena. 7. Geranium Bijou. 8. G. Cottage Maid. 9, 9. G. Triomphe de Paris. 10. G. Flower of the Day. 11. G. Trentham Rose. 12, 12. G. Rubens Improved. 13. G. Stella. 14, 14. G. Pink Nosegay. 15. G. Bijou. 16. G. Cybister. 17, 17. Calceolaria Aurea floribunda and scarlet verbena. 18. Geranium Diadem. 19, 19. G. Alma. 20. G. Sydonie. 21. G. Duchess of Sutherland. 22. G. Diadematum. 23. G. Vitifolia. 24. G. Diadematum erubescens. 25. G. Addisonii. 26. G. Quercifolia coccinea. 27. G. Golden Fleece. 29. G. Madame Vaucher. 30. G. Cloth of Gold. 31. G. Chancellor. 32. G. White Perfection. 33. G. Eve. 34. G. Magenta. 35. G. Alfred. 36. G. Rose Queen. 37. G. Beauty of Blackheath. 38. Ribbon border: 1st row, Dactylis glomerata variegata; 2nd row, Tropæolum Sparkler; 3rd row, Geranium Christine; 4th row, Geranium Stella.

SCALE OF FEET

The plans which follow, on pages 24, 25, and 26, will, it is hoped, be clearly understood by the aid of the accompanying enumeration of the plants employed in furnishing them.

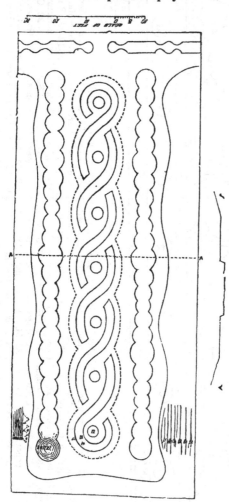

1. Mottled Dahlia. 2. Ageratum mexicanum. 3. Geranium Lady Middleton. 4. Calceolaria Aurea floribunda and C. Amplexicaulis, mixed. 5. Tropæolum Sparkler. 6. Geranium Bijou. 7. and 8. Vandyke of Cerastium and Blue Iobelia. 9. Edging of Cerastium tomentosum. 10. Verbena Purple King. 11. Scarlet Geranium. 12. Yellow Calceolaria. 13. Cerastium tomentosum. 14. Verbenas, mixed scarlet and white. 15. Viola montana var. 16. Geranium Manglesi. 17. Verbena Scarlet. 18. Calceolaria Prince of Orange. 19. Geranium Punch and Waltoniensis. 20. Chrysanthemum pinnatifidum. 21. Geranium Rubens. 22. Dahlia Beauté de Massifs.

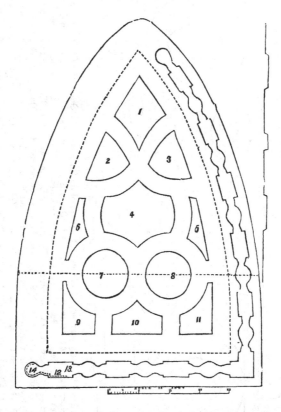

1. Geranium Christine, and margin of 18 inches Perilla. 2. Geranium Stella, and margin of 24 inches of Yellow Tom Thumb Tropæolum. 3. Centaurea ragusina and Amaranthus tricolor. 4. Geranium Bijou, and edging of 24 inches Verbena Purple King. 5. Verbena venosa and Viola lutea, mixed. 6. Geranium Gold Leaf, and edging of 12 inches Geranium Little David. 7. Geranium Lord Palmerston, and edging of 18 inches Tropæolum nanum punctatum (yellow with scarlet spot). 8. Geranium Diadematum, and edging of 18 inches Gnaphalium lanatum. 9. Geranium Queen of Queens, and edging of 18 inches Viola montana var. 10. Tropæolum Stamfordianum, and edging of 24 inches Verbena Ariosto. 11. Geranium Silver Queen, and edging of 18 inches Lobelia Paxtoniana. 12. Edging of Cerastium tomentosum. 13. Verbena pulchella and Geranium Golden Chain, mixed. 14. Lobelia Paxtoniana and variegated Alyssum mixed.

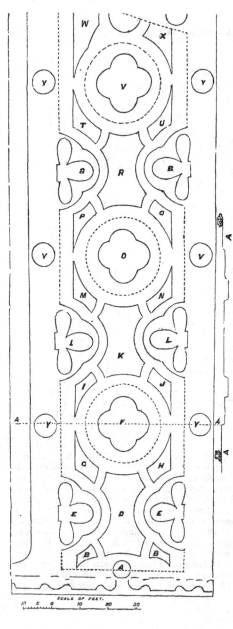

On either side of this a ribbon-border differently planted, as follows :—1. Tropæolum Sparkler.   2. Geranium Christine.   3. Calceolaria Amplexicaulis and Aurea floribunda.   4. Ageratum mexicanum —1. Dactylis glomerata variegata.   2. Geranium Christine. 3. Calceolaria Amplexicaulis and Aurea floribunda.   4. Geranium Stella.   A. Purple-leaved Clover and Chamæpuce diacantha.   B. Geranium Alma.   C. Geranium Countess of Warwick. D. Geranium Christine.   EE. Geranium Tom Thumb and Calceolaria canariensis, mixed.   F. Geranium Stella.   G and H. Geranium Madame Vaucher.   I. Geranium Little David.   J. Geranium Spread Eagle.   K. Geranium Flower of the Day.   LL. Calceolaria Victor Emmanuel and Verbena Ariosto, mixed.   M. Geranium Vivid. N. Geranium Triomphe de Paris.   O. Geranium Cottage Maid.   P. Geranium Helen Lindsay.   Q. Geranium Princess of Prussia. R. Geranium Bijou.   SS. Geranium Tom Thumb and Calceolaria Aurea floribunda, mixed.   T. Geranium Pink Nosegay.   U. Geranium Cybister.   V. Geranium Trentham Rose.   W. Geranium Hendersoni (variegated).   X. Geranium Argenteum (variegated). YYYYYY. Blocks of various new Geraniums on trial.

The next design, p. 28, will explain the principles of Parquet colouring. It is the work of a talented head gardener in a large private establishment, who says of it : —
" The planting and condition of this bed had many admirers. I do not claim any particular merit for it, although it was my own work; because the position of the bed, and the restrictions under which that position place the planter, can only be known to those most interested in it. Still, the fact that many wish to copy it speaks well for the principles observed, and of the probability that it is worthy of imitation in other places. It is necessary to state that it was formed out of the middle of a broad stone terrace adjoining the mansion. But a new wing being added to the house, and the principal window of that wing looking down upon the stone terrace, it was considered desirable to break up the monotony of the stone terrace by forming this parquet garden in the centre. The restrictions imposed upon the planter will be evident to the reader. As the principal windows are in the new wing, it is from that point from which the bed is viewed, and it should show all its features from that point without being distasteful to the eye when looked upon from any other. It was the opinion of many that the same plants and the same arrangement would look well as a double ribbon border in any position. The splashes of yellow introduced at the corners, cutting off, as it were, the sharp angles of the lines, were objected to; but, for my own part, I never regretted that feature, because it broke up the stiffness of the arrangement, and it softened down the tones of the massive lines of Stella geranium. As a rule, angular arrangements are objectionable, but, as they must be sometimes adopted, this example may be useful in your series of bedding examples."

The next examples, p. 29, represent the embellishments of an entrance court, which is remarkably well-kept, being richly stocked with coniferous trees, and the walls densely clothed with the choicest ivies. The central walk is flanked on each side with small grass-plots, on which are marked out oblong compartments and narrow scrolls as in the small figure. The ground-work is wholly formed of statuary marble broken to the size of hazel-nuts, and laid down on a bottom of concrete, to prevent the soiling of the white marble by wormcasts. The scroll is, therefore, produced in relief on a snow-white ground, and is planted thus:—A A, Golden Fleece

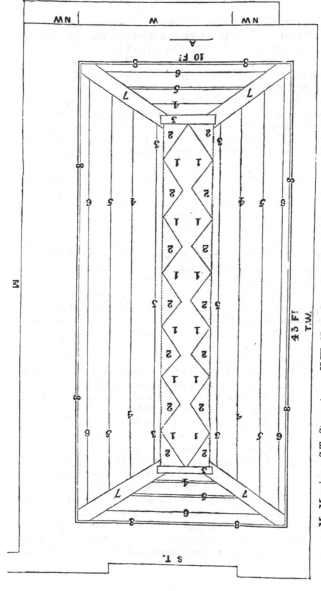

M. Mansion. S T. Stone steps. N W. New wing. W. Window, looking south. T W. Terrace wall.
1. Perilla nankinensis. 2. Cineraria maritima. 3. Geranium Stella. 4. Geranium Madame Vaucher. 5. Geranium Golden Chain, the flowers picked off. 6. Viola cornuta. 7. Yellow Calceolaria. 8. Stone Kerb.

Geranium; BB, blue Lobelia; C, Alma Geranium. This scheme is admirably adapted for small gardens and entrances, and requires only ordinary skill to work it out successfully. Immediately in front of the gate is a circle of grass turf, with standard bay tree in the

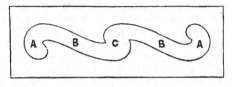

centre, and four equidistant horns or compartments, as represented in the figure. The planting of this design is very

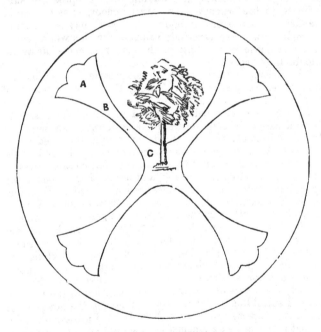

simple, but most effective. A is a bold clump of Perilla, forming the termination or mouth of the horn; B, Flower of the Day Geranium, which is continued to the centre C, so that the design has but two prominent colours—purple-bronze and creamy-grey, a subordinate colour being the cerise blossoms of the geranium. The effect of this style of planting is enhanced by the rich foliage of the evergreen shrubs in the adjoining borders, and its intense brightness is relieved by the pleasant view of a large and rich green garden beyond.

Space need not be occupied with ribbons and scrolls, because of their simplicity in the first place ; and because, in the second place, the larger schemes include all the smaller ones, so far as principles and details are concerned. It may be remarked, however, for general guidance, that scrolls and ribbons must always be decisive in colouring, sharp and bright, and either strictly linear in arrangement, or so arranged that the vandykes, crescents, and waving lines adopted are subordinate to the primary linear arrangement, so as to sustain the idea of a scroll or a ribbon as the case may be. Plants of small growth are especially valuable for this work, which should be dense in planting of the very best materials available for the purpose.

A curious and eminently pleasing style of massing has lately been adopted by Mr. Mason, the superintendent of Princes Park, Liverpool. This is known as Tessellated colouring, the colours being repeated in small blotches, with sharp dividing lines to separate the groups, like a series of dotted ribbons placed side by side to form a connected piece. In this system, foliage plants are freely employed side by side with flowering plants, and the result is a rich mosaic or tessellated pattern extremely pleasing and interesting to look down upon, but wanting in decidedness when viewed from a distance. The examples figured occur on large breadths of green turf, which greatly aids the general effect ; in fact, gravels of any kind would be unsuitable for a groundwork, by the too near approximation of their colours to some of the oft-repeated neutral tints in the planting. It is of great importance to select for the purpose plants that are likely to continue good throughout the season, for a failure anywhere would be particularly disastrous on account of its repetition in the form of a broad sprinkling of blank spaces amongst the flowers. It is not less important either to select plants of the same height, or that admit of being pinched back should any of them overtop their neighbours. The schemes are explained in the enumeration of subjects employed in producing them. The blank lines are planted the same as those they correspond with.

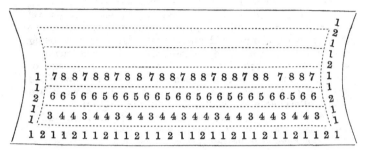

Outside Row: 1, 1. Lobelia speciosa; 2. Golden Chain Geranium, dwarf bushy plants. *Second Row:* 3. Geranium Miss Kingsbury, with flowers picked off; 4, 4. Geranium Little David. *Third Row:* 5. Calceolaria aurea floribunda; 6, 6. Dark-leaved Beet. *Centre Row:* 7. Centaurea ragusina; 8, 8. Pink Geranium (seedling).

Scale, 4 feet to the inch.

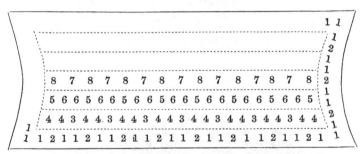

First Row: 1, 1. Dactylis glomerata variegata; 2. Geranium Little David. *Second Row:* 3. Dark-leaved Beet; 4, 4. Calceolaria aurea floribunda. *Third Row:* 5. Geranium Waltham Seedling; 6, 6. Geranium Bijou, with flowers picked off. *Centre Row:* 7. Perilla; 8. Large plants of Centaurea ragusina.

Scale, 4 feet to the inch.

A few words on leaf embroidery must suffice to close this chapter. An interesting and extremely beautiful example of this style of dressing was represented in a coloured plate in the FLORAL WORLD for March, 1871. The reader, who can refer, will observe that it is equally well adapted for the grandest terrace garden, or the quite humble and unpretending grass plot in a villa garden. It may be likened, in a general way, to a hearthrug or Turkey carpet pattern, though of course it must be less complicated, and the materials employed being chiefly leaf colours, blend in the same soft, warm manner, with more thorough distinctness in the several blocks of colour, because there are no green leaves to interfere with the unity of each. The most useful plants for this work are coleus, alternantheras, the "golden feather," pyrethrum, centaureas, iresines, perillas, amaranthus melancholicus, and a few of the more distinct echeverias and sempervivums. It is the latest novelty in flower garden embellishment, but is destined, we cannot doubt, to become extremely popular because of its richness; the comparative ease and certainty with which satisfactory results may be obtained; the long continuance, without change, of the colouring produced in the first instance; and the oneness of colour in each separate line or block, the whole scheme improving as the season advances, without the possibility of the occurrence of those blanks we are accustomed to in the case of flowering plants, which are usually without flowers when planted, and are apt to go out of flower for a week or two in the very height of the season.

CHAPTER III.

THE " bedding system," as commonly understood, is an idea only half developed. It is very much to be feared it will never be known as a complete system, but that it is doomed to remain an example of arrested development, so far as the mass of the people are concerned. Let us consider for a moment the case of a geometric garden occupying a conspicuous position, and intended as one of the principal, perhaps *the* principal embellishment of the garden. As a design in black earth, and green box, and grey gravel, its merits are not worth considering; but we are always prepared to consider its merits in connection with its purpose, and will pass judgment upon it when filled with flowers, just as we would prefer to judge a picture-frame with the picture in it. Well, we will wait until the month of May. By the end of that merry month of flowers the beds are all filled; but the plants are puny bits of things, and must have time to "make themselves." So we will wait until June. By about Midsummer-day, a pretty fair sprinkling of flowers will be seen in the geometric garden, and we may then make an estimate of its artistic value as a design, as well as of the skill employed in planting it. From Midsummer-day to Michaelmas-day, when usually the first autumnal frost occurs, the best of the summer bedders are extremely gay. For just three months, in fact, a few days more or less, according to the season, the parterre planted agreeably to custom is brilliant in the extreme, and for the remaining nine months of the year it is a dreary blank. It is like a display of fireworks, glorious while it lasts, but "ere we can say, 'Behold how beautiful!' 'tis gone," and the darkness that follows is rendered more profound by contrast with the light that dazzled us. Yet, for the sake of this temporary glory, ten thousand gardens, that

3

would otherwise have been rich in attractions of a permanent character, and comparatively exhaustless in interest, have been reduced to the condition of manufactories, and the summer show, as a proof to all observers of what the factory could produce, has been considered sufficient return for the sacrifice of all that should make a garden at once a training-ground for mind and morals, and a recreative feature of the house itself.

The bedding system has its uses as well as its abuses. In many a place it operates injuriously, by contracting the ideas of those who profess to love their gardens, and absorbing energies and appliances for the accomplishment of paltry results, which might be devoted to purposes conducive to the production of a really enjoyable garden. But for its own particular purpose, and in its proper place, with liberal sur-roundings, and with means for its proper vindication, the bedding is not only invaluable in its present imperfect state of development, but worthy of all the energy and thought it may demand for its completion. Its one grand defect admits of the most perfect remedy, but every step in the remedial process is attended with expense and labour. To be sure, it is not possible to have a display of flowers in open beds the whole year round, but there may be four displays of some kind in the course of twelve months. From March to May, the parterre should present a succession of masses and lines of spring flowers ; say crocuses, tulips, forget-me-nots, scillas, iberis, alyssums, pyrethrums, pansies, daisies, and polyanthuses. Then should follow the summer display of geraniums, verbenas, petunias, and the rest of the generally accepted furniture. At the instant of these declining in beauty, early-flowering pompone chrysanthemums, brought in from the reserve ground, might be planted in their places, to make a brilliant display from the middle of October to the middle of November. Then the spring display must be prepared for by planting bulbs and herbaceous plants, and a few beds, and centres of beds, might be left wholly or partially vacant in this planting, in order to be filled with showy evergreen shrubs carefully lifted from the reserve ground, or grown in pots for the purpose and plunged. The programme here sketched out is not strictly like the blind man's fiddle that he made " out of his own head," for the author has carried it into effect and kept it going for years, and has thus tested and tried all its capabilities and difficulties. In all well-kept gardens, the parterre should be planted at least twice a year, namely, in May for the summer display, and in October for the spring display. The employment of chrysanthemums for autumn, and ever-green shrubs for winter, demands much space, makes much labour, and needs very nice management, whether the system of planting or plunging be resorted to for the sake of con-tinuous enjoyment of "a gay garden." Considering a dis-play of spring flowers to be absolutely necessary, it will be proper to offer a few practical remarks on the course to be pursued by those who would secure it at the least possible cost, and with the best possible result. Having disposed of that part of the subject, the summer display will demand attention.

The most useful materials for a display of spring flowers are to be found amongst the hardy bulbs. The kinds on which we must chiefly depend for the principal effects out of doors are the crocus, snowdrop, tulip, and hyacinth. Where required to be used in large quantities these may be had in distinct and striking colours, and of good quality, at very cheap rates. It is most important for people who really wish to do the best with their gardens, to know that a show of spring flowers does not necessitate extravagant outlay ; for though we may spend five-and-twenty pounds upon a single tulip, and five pounds more to grow it properly, as good an effect may be produced, if the embellishment of the garden is all that is required, by a bulb costing one penny, and an additional farthing for the expense of cultivation, inclusive of labour, manure, and rent. Nevertheless, there is a popular dread of bulbs for use on a large scale as ruinously expensive.

There is also another difficulty, and that is, that gardeners wish to deal with them as with summer bedders. The latter they dispose so that all shall be in bloom at the same time, and they want to do the same with a collection of bulbs, but Nature is against them. It is a very easy matter indeed to plant the several sorts of bulbs so that their blooming at different times is a positive advantage, whether in continuous borders or in beds that constitute groups all under the eye at the same time. For instance, in a geometrical garden laid out on a lawn within view of the drawing-room windows, all the beds that correspond with each other in the pattern can be planted with the same kinds of bulbs, so that when

these are in bloom there will be the same harmony of arrangement as if the beds were the same throughout. A simple
scheme will make this plain : suppose a set of eleven angular
beds on a lawn as here represented, the gardener's object may
be to have several kinds of bulbs in bloom all at the same
time, and that is just the very thing that cannot be accomplished. But for months together there may be abundance
of flowers in rich masses, without any lop-sided anomalies, as
the planting of the beds will show :—

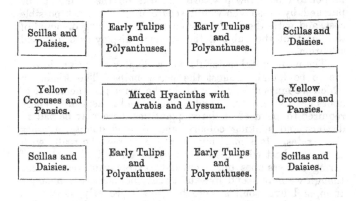

It will be seen that it matters not whether the various
plants employed bloom altogether or in succession, each
separate class will be in bloom in its own season, and yellow
crocus on one side will have a match in yellow crocus on the
other, and the same with all the rest. But this simple scheme
may be improved by using all the smaller bulbs as edgings
to the larger beds. Suppose them all edged with snowdrops,
then early in the year the whole scheme will be gay with
white flowers. Next the snowdrops plant crocuses, and as
the snowdrops go out of bloom these will succeed them ;
then as the crocuses decline, the hyacinths and tulips, forming the principal masses, will come to their full splendour,
and the season of spring flowers will be prolonged almost
to the time for turning out summer bedders. There are
numbers of early-flowering herbaceous plants suitable to
plant with the bulbs to make masses of verdure all the

winter, and a rich surfacing of flowers in the spring, and at
the proper time the beds should be cleared of these and the
bulbs together, for the customary summer planting.

The supposed expensiveness of bulbs, that deters people
from using them largely, is a most injurious fallacy, for they
are by no means so costly as supposed. But there is another
impediment, and that is the supposition that the soil must
be prepared in some mysterious manner with elaborate composts, and processes which few understand. Now the simple
truth is, that for all the bulbs and herbaceous plants commonly used for masses in the flower garden, the only preparation necessary is to break up the ground well and manure
it moderately, leave it a few days to settle, and then plant.
If the soil is wet it must be drained ; but that is necessary
for everything else cultivated in it. Scarcely anything worth
having will grow in ground where the drainage is not either
naturally or artificially sufficient to remove surplus water
quickly, so that the soil is never more than reasonably moist.
All the bulbous-rooted plants like a rich sandy soil, but there
is no occasion for composts, and all tedious operations are
unnecessary.

Now as to the cost. All the best bedding tulips may be
obtained at from five shillings to nine shillings per hundred ;
and the most expensive kinds will never cost more than four
shillings per dozen. A reference to any of the bulb catalogues
will show that if good colours are the desiderata without
reference to the peculiar excellence of varieties delicately
striped or finely formed, a few pounds will go a long way to
make the garden an agreeable attachment to the house during
the early months of the year, instead of, as it too often is at
that season, a dreary wilderness. In all the bulb catalogues
"mixtures" are advertised at a cheap rate. When these
mixtures are in *distinct colours* they may be very useful for
those who are obliged to make the most of a small outlay. But
mixtures of colours are objectionable in geometric arrangements, and in this scheme we should admit only one mixture,
and that would be of hyacinths. If we were to plant a set of
beds like those in the scheme above, we would have the
edgings of snowdrops and crocuses all through. The two
crocus beds we would also edge, by planting blue crocus
inside the line of the snowdrops ; all the rest of the beds we
should make the second line of yellow crocus. The four

corner beds we should plant solid with Scilla siberica. The
four tulip beds should be of four kinds only, the bulbs five
inches apart all over, and the hyacinths mixed thus :—

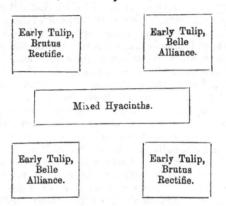

Early Tulip,
Brutus
Rectifie.

Early Tulip,
Belle
Alliance.

Mixed Hyacinths.

Early Tulip,
Belle
Alliance.

Early Tulip,
Brutus
Rectifie.

The principles which govern the use of bulbs in solid masses
do not strictly apply to their use in borders. Here they can
be used in close lines as ribbons, or in distinct clumps, which
are better than lines certainly. Compare a line of snowdrops
or crocuses with a set of clumps, and the latter will always
be pronounced the best disposition of them. As the different
kinds of bulbs bloom at different periods, there will be the
same succession as in beds, and the places for each will be
determined by height only—say for front line clumps of snowdrops and Scilla siberica, nine inches apart all through ;
behind that front row clumps of yellow crocus ; behind that
again, clumps of blue and white crocus, not mixed, but
distinct and alternating ; then hyacinths, and for the back
row early tulips.

With the exception of hyacinths, all the bulbs we have
named will increase in value every year if planted in a sound,
well-drained, well-manured soil, and the more sandy the soil
the better. They should be planted before they have grown
much, and be taken up when the foliage is decaying, and be
laid in some shady place covered with a little mould to ripen
before being stored. Crocuses and snowdrops need not be
removed every year, but once in three years. They should

be taken up, the ground should then be trenched and manured, and the bulbs planted again. Borders appropriated to a dis-play of these in spring may be sown over with annuals with-out injury to the bulbs, and to render a yearly lifting of them unnecessary. As to hyacinths, they rapidly deteriorate unless subjected to careful systematic cultivation. As a rule, the best plan is to purchase a fresh supply every year, and throw away those that have flowered. The other kinds of bulbs do not deteriorate if carefully managed as bedding plants. One more remark may be worth making, it is that all the most valued bulbous and tuberous rooted plants thrive amazingly well in the smoky atmosphere of great towns.

We will now briefly indicate a few of the more important points that require consideration in connection with the sum-mer bedders. To begin with, we must divide these into two classes—1, those that produce effect by their flowers; 2, those that produce effect by their leaves. In the first section the most important plants are Verbenas, Petunias, Calceo-larias, Lobelias, Lantanas, Heliotropes, and Tropæolums. In the second section the most valuable subjects are Coleus, Amaranthuses, Alternantheras, Iresines, Perillas, Centaureas, Cerastiums, Gnaphaliums, Pyrethrums. From the great family of Geraniums (zonate pelargoniums) we can select plants for both classes, and so far as they serve the purposes required by the flower and leaf colours, they are without question the most useful bedding plants in cultivation. But the question arises what constitutes a bedding plant? Before attempting an answer to this question, it must be remarked that although such noble subjects as Agaves, Yuccas, Cannas, Humeas, and Beaucarneas, may be employed to enrich the parterre, our chief concern now is with the plants employed in *flat* colouring, for such are the bedders proper. This con-sideration suggests the limits within which we may select plants for bedding. They must be decisive in the colour of leaf, or flower, or both; they must be of comparatively dwarf habit, or admit of being trained close to the ground to pro-duce the same effect as dwarf plants; and they must present the appearance which renders them valuable as agents in colouring for a considerable length of time, and the longer the better. Other qualities we need not make note of. It is evident that a plant selected for its flowers will prove but a poor bedder, if those flowers are presented in a succession of

efforts with considerable intervals between. Hence, antirrhinums and pentstemons, which, as garden flowers, are most
beautiful, cannot be considered valuable bedders. Pyrethrums
flower too early for the summer display, and phloxes flower
too late. A vast enumeration might be made of plants commonly regarded as bedders, that really do not belong to the
category ; but it is sufficient to say that given all other needful qualities, *continuity* of effect, whether by leaves or flowers,
is an indispensable quality. Here we light upon an interesting distinction between such as we may call flowering plants,
and such as we may called leaf plants. Under the best of
circumstances, we must wait for the first, for even if we plant
them in full bloom, the change of conditions consequent on
planting will soon cause them to cast their flowers, and some
time must elapse ere they produce a succession. With leaf
plants, the case is quite different. They show their colour
from the first, however weak it may be, owing to the smallness
of the plants ; and they improve every day. With flowering
plants, the first display is of green leaf, with accidental dots
of colour. With leaf plants, the first display is the same as
the last, save and except as to intensity. If a verdict as to
relative values must be given here, the leaf plants must have
it certainly, and the latest fashion in leaf embroidery will
amply justify this preference of leaves to flowers for colouring
of the richest and most artistic character. As, however, it
will be long ere the leaves drive out the flowers, the last must
have attention in these pages, without regard to their possible
eclipse in years to come, or the great probability that, after
all, the flowers may be in the end triumphant.

The shortest and simplest way of making a display of
bedding plants is to buy them when wanted, and dig them
into the beds as manure when the autumnal rains have spoiled
their beauty. And it needs to be said that this is not so
extravagant a mode of procedure as it appears. To be sure,
the plants will cost money, and the outlay must be repeated
every season, so long as bedders are required. But those
who raise their own plants, and keep up a stock for bedding,
do not obtain their results for nothing. They must employ
skilled labour, and make use of glass, and burn fuel, and
occupy space for a mere manufacturing business which, if
judged from any high standpoint, offers but little to interest
the enthusiastic horticulturist. The system of purchasing

and destroying will suit those who have no glass, and it
may suit many who have, because, by setting free labour,
glass, fuel, and ground-rent from the production of bedding
plants, the means may be found to grow pines, grapes,
mushrooms, melons, the very noblest stove and greenhouse
plants, and many grand conservatory plants which are quite
unknown to those who find in geraniums and calceolarias
the sole objects of horticultural care, and the only worthy
subjects of horticultural enthusiasm. It is quite a question
whether thousands who grow their own bedding plants do
not pay more for them than those who purchase annually.
However, it is our business to help both parties, and we close
this paragraph by remarking that, without a sufficiency of
glass, it is next to impossible to carry out the bedding system
with home-grown plants, and a regular routine of cultivation
must be followed to enable the planter, in the month of May,
to fill the parterre according to the arrangement predeter-
mined on. It happens, fortunately, that a few simple directions
on cultivation will apply to nearly all the bedders enumerated
above, and these may very properly be presented in the next
chapter.

It remains now, to complete this section, that a few
remarks of a general kind should be offered on the distribu-
tion, and proportions, and relations of the colours employed
in the furnishing of the parterre. The reader will not need
to be informed that a tasteful display can only be obtained
by a judicious employment of the materials at the command
of the planter. They may be sufficient for the production of the
most artistic effects, and yet may be made subservient to mere
vulgarity, or to a meaningless expression of weak harmony,
unless they are proportioned and disposed with skill. Let us,
therefore, consider the whole case in a comprehensive manner,
though a few of our remarks may be but amplifications of
points already succinctly stated.

All our general views of Nature afford us hints of the
laws by which the disposition of colours should be regulated.
But particular views are still more instructive to the artist.
Let us behold the meadows in the month of May, and rejoice
in the golden glow of buttercup blossoms with which they
are overspread. What does the sight consist of? You will
be disposed to answer, perhaps, that it consists of a green
groundwork covered with dottings of yellow. And you are

right therein; but it will be observed that in the foreground
the green is of great breadth, and some intensity—that, in
fact, it is a more distinctive feature than the yellow. But
look at the mid-distance. There the green groundwork is
subdued in tone, and the yellow has gained in strength, so
much so that the green is almost overpowered by it, and we
call it—properly too—"the field of the cloth of gold." But
now, observe the background. If the field extends to a sufficient distance from the eye, its farthest boundary is a sharp
bright line of gold; the green groundwork is lost altogether;
the buttercups, which near our feet are scattered so that
between every two or three tufts of flowers there are distinct
hummocks of grass, are in the far distance packed so close as
to present to the eye a solid golden band reaching across the
field, and which, if there is a copse or a heath beyond, looks
all the brighter and sharper by contrast. That these different
appearances of the field are delusions, need not be explained.
We see the distant buttercups at a lower angle than those
that are near, and the gradual strengthening of the yellow,
and weakening of the green, as the eye ranges across the field,
are phenomena resulting solely from the different angles at
which each successive distance is viewed: we look *down into*
the grass at our feet; we look *along the surface* of the whole
vegetation as we glance to the distant parts of the scene, and
the horizontal line of vision passes through all the buttercups,
and does not touch the grass at all.

We can agree on two points—first, that the change from
a predominance of green to a predominance of yellow is perfectly natural and easily understood; and, second, that it
affords immense delight to the eye—so much delight, indeed,
that the most fastidious colourist amongst us could scarcely
wish for a finer effect than is every year produced by every
meadow that is well sprinkled with buttercups. Now, what
is the idea of that scene when considered artistically? The
idea is, that one colour may dominate, may make other
colours subservient to it, and so afford pleasure to the eye.
We have a hint here of the value of what may be called
dominant colouring, and which in bedding displays may be
worked out to grand results. Let us suppose we have to
colour a group of panel beds, or a geometric scheme on a
terrace. By selecting one strong colour to determine the tone
of the whole group, we secure, in the first instance, an idea;

that idea makes itself expressed in a feature; and the result
will be, gratification of a higher order than would result
from a disposition of colours without regard to any principle
at all. It will, of course, greatly depend upon the nature of
the design to be painted, the nature of the surroundings, the
degree of grandeur of the buildings, walks, lawns, and so
forth, how this idea is to be applied; but an artist in colour
will not be long in determining anywhere. Probably in nine-
tenths of all the private gardens, the best colour to take the
lead all through a complete scheme would be scarlet. But it
does not follow that, if we select scarlet, we are to use no
other colours. Nothing of the sort. It is to be understood,
in such a case, that scarlet is to rule; that there are to be
several shades of scarlet lending aid to each other; and that
other colours are to come in as dividing lines, separating
blocks, boundaries, and relief agents—all these being so used
as to lead the eye to scarlet, and again to scarlet, claiming
for themselves no importance whatever.

Suppose, for the sake of illustration, we proceed to plant
a group of beds, beyond which there is an enclosing ribbon
border. We may have in the centre a neutral tint for the
purpose of helping the eye to range over the whole design
without being drawn to the centre by any undue attraction.
If we make the centre yellow, we ruin the scheme; the eye is
drawn to it—fixed, and charmed, and spell-bound by it; it
wars against the predominance of scarlet, and the idea with
which we begun is already trodden under foot. But we may
have there a variegated-leaved geranium, and one of the
creamy section will be preferable to a white leaf. If more
colour than a creamy leaf variegate would afford were required, some soft shade of scarlet, or red, or pink, would
answer admirably; and a reddish-lavender, puce, or rose,
would be admissible. The outer beds all through, in which
the great leading features of the design are made manifest,
should be in the strongest tones of scarlet; and, if the
pattern has some complicated fillings up, relief colours will
be wanted in them—not for the purpose of introducing as
many colours as possible, but solely to help out the expression
of the whole, and give the scarlet its full importance as the
one colour to which every other is subservient. For intermediate dots and relief-agents, however, rosy-purple, yellow,
white, and even blue, will be admissible; but the purple will be,

most suitable for relief where any considerable breadth of
colour is wanted, to separate large blocks of scarlet, or to fill
any odd portion of a design which, like the nose on the face,
has no relation to any other corresponding feature, but serves
to separate and systematize them. For distinct dots, blots,
angles, and small fillings-in, yellows and blues will be in good
taste; and, if well used, will help to bring out the good purpose of the design.

For the edgings of the beds there will be an admirable
opportunity for the free use of gold and silver leaves. Sharp
white, creamy, or amber lines bordering the whole will be in
better taste than edgings of all colours; yet, in a group of
scarlet beds, two or three styles of edging are admissible, if
very distinctly arranged, so as to balance every part of the
design; and, of course, the larger the scheme, the more
various may the edgings be, and on them will depend in a
large measure the picking out of the design, so as to enable
the eye to apprehend and appreciate it from some point of
view fairly commanding the whole.

To colour the enclosing ribbon in the same way as the
beds would show a poverty of invention, not a deficiency of
taste. Scarlets with relief agents are perfectly admissible
there; but tones of red and purple, and an outside edging
differing altogether in character from the edging in the beds,
will be preferable. We will suppose that the only yellow in
the beds occurs in the form of small dots, and is therefore
inconspicuous. For that very reason, a yellow edging to the
enclosing ribbon will be quite appropriate. It would be the
reproduction of the golden fillet which the Assyrians and
Egyptians used so successfully in their bold red colourings.
For the second line, say three shades of red; then a bold,
sharp line of white, grey, or pale blue. If the breadth of the
ribbon needed more lines, three shades of red might be employed again, with a finishing line of something bold enough
to make a definite boundary.

It is not pretended that there is anything new in this
method of colouring. There is nothing new under the sun.
It has been Nature's mode of procedure ever since the first
day of creation. The earth has shown to man, for his delight,
successive breadths of dominant colours—white, primrose,
and yellow in the spring and early summer; orange, red, and
crimson in the full summer; russet, brown, bronze, and pur-

ple in autumn. Look at the heathy hills in July and August,
and how do they compel all the dots of green, and red, and
white, in the adjoining meads and hedgerows, to become subservient to their own vast and wonderful sheets of crimson,
which the ling then clothes them with, as with a garment of
fire! Or look, in spring-time, at some of the moist grass
lands of the southern counties, when the lady's-smock is in
bloom, and how the snow-white vesture takes to itself a stripe
of green as a girdle, and a sprinkling of yellow globe flowers
as gold tassels and trimmings, the white still predominating,
and by that fact making a deep and joyful impression on the
mind of the beholder. Nor is it in any case hard to carry
the idea into effect in planting; it is, indeed, most easy. It
makes the routine of bedding more simple than when it is
inspired by untaught fancy, and does away with all those
difficulties that beset mosaic painting, where one of the chief
objects is to establish a balance of all the colours.

> " Light and air
> Are ministers of gladness; where these spread
> Beauty abides and joy : wherever life is
> There is no melancholy."

CHAPTER IV.

CULTIVATION OF BEDDING PLANTS.

ONE of the principal reasons why, in many instances, bedding plants are slow in making their proper effect, is that they are preserved during winter in greenhouses constructed for better purposes. A house suited for camellias, azaleas, and heaths, will not suit bedding plants well, unless they are placed on shelves very near to the glass. Abundance of light all the winter long is one of the most important conditions, for if the plants are far removed from the glass, they become attenuated, make long weak shoots, and suffer considerably when planted out, no matter how favourable at the time the weather may be. Moreover, when housed with proper greenhouse plants, they are generally kept too moist and too warm, and the result is that they grow when they ought to rest, and are in a tender state when the time arrives for planting them. The best place for the hardier kinds of bedding plants, such as geraniums, petunias, and verbenas, is a well-built brick pit or greenhouse, with very low roof, in which the plants can always be kept very near the glass, and the management of which, as to temperature, moisture, and air, will be considered with reference to the bedding plants, and not with reference to other things that may be mixed with them. To describe plant houses in detail is no part of the purpose of this book; but it is necessary to the completeness of a practical consideration of bedding plants, to offer examples of houses adapted for their preservation. The first shall be the simplest, and the cheapest possible—a good useful pit, costing five or six pounds at the utmost. The walls are four-inch brickwork; and in order to make them more secure against frost, as well as improve their appearance, a bank of earth, one foot wide at the base, and sloping upwards to the sill, might be thrown against them, and neatly turfed over. The furnace is sunk below the ground level, in a pit at the end, as indicated by the dotted lines, which communicates with a flue running along the inside of the front wall to the chimney. A movable brick should be let in at *a* and *b*, for the purpose of cleaning the flue. A common furnace, such as is used for a small copper, will do; and the furnace-pit should be covered with a

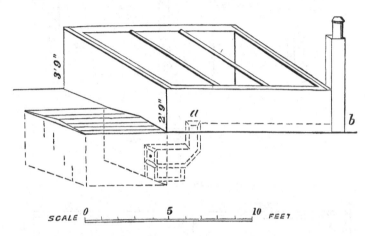

SCALE 0 — 5 — 10 FEET

folding lid. On the top of the pit-walls is a wood sill, 4½ inches by 2½ inches, and cross-bars to slide the lights upon; the whole covered with three well-glazed lights. The plant stage inside the pit may consist of simple boards, which can be raised or lowered, according to the wants of the plants, by placing them on blocks of wood.

The figures on page 48 represent a suitable pit designed in detail in order to simplify the labour of production. It is estimated that the whole cost should not exceed £25, and, when carried out, would form a complete *multum in parvo* for the gardening amateur. Fig. 1 represents the ground-plan and section of warm-pit, in the back of which is a path, *k*, a bed for tan or leaves, *a*, in which roses, lilacs, azaleas, rhododendrons as well as a supply of hyacinths, narcissus, etc., may be forced during the dull months of winter; and in summer, achimenes, gloxinias, and many of our finest stove-plants may be grown, as well as a few pots of strawberries on

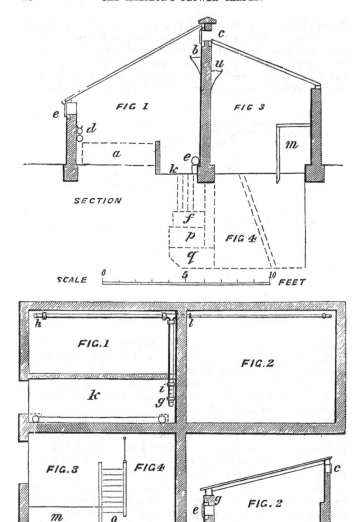

shelf *b*. The roof is a fixture, but ventilation is secured by three openings back and front, *c*, each 18 inches by 9 inches, over which slide boards in a groove. The boards are connected together by means of a stout wire, running from one to the other, with a handle at the end, so that all may be opened or shut at once by merely pulling or pushing the handle. The ends of this part may be either all brickwork, or the front wall returned; and above that may be glass, according to the taste of the builder. The latter would be the best-looking plan, but would cost a trifle more than brickwork. Atmospheric heat is obtained from two four-inch hot-water pipes *d*, the flow rising at *g*, and the return descending to boiler at *i*, and flue formed with nine-inch drain-pipes. Wherever an elbow occurs in this kind of flue, it is well to use a few bricks, covering with a pavement, the removal of which at any time will enable a flue-brush to be got in for cleansing the flue. It is also to be remembered that a flue always acts best when the furnace is sunk considerably lower than the line the flue traverses, otherwise the air stagnates in it, and causes the smoke to rush out at the furnace-door. For the heating, a very small boiler will do. There will be 36 feet of four-inch pipe, two elbows, one syphon, and a supply-cistern, 9 inches square, for fixing at *h*, required for Fig. 1; also, two diminishing T-pieces, one stop-valve *l*, one two-inch syphon, and 18 feet of two-inch pipe, for Fig. 2; a furnace-front and bars, and small soot-doors placed opposite the principal flues for convenience of cleaning from soot.

For the building must be provided 4000 red bricks, 250 white bricks for floors, 10 feet of coping-bricks, one chaldron, or 36 bushels, of lime, and three loads of sand, and 20 feet of 9-inch drain-pipe for flue and chimney.

Fifty-four feet of wall-plate, 4½ inches by 3 inches, for the various roofs to rest upon; and if the ends of the pit, Fig. 1, be only bricked up as high as the front-wall, and the rest part glass, about 14 feet more will be required; also, for the jambs and lentels for two doors, 34 feet of the same scantling, making about 102 feet.

For the roof of Fig. 1, 180 feet of sash bars, at 8 inches apart, will be required; and about 40 feet of 3 by 4½-inch scantling, to lay into the walls as bond-timber for the pitch of the roof and ventilators. About 32 feet of 1 by 9-inch board for shelves *b*, *u*, and ventilators, *c*, five iron brackets,

ditto, a few feet of spline for ventilators, and ⅜-inch iron rod
for the same; a ladder *o* for stoke-hole, one door and threshold for potting-shed, Fig. 3; also, one door, partly glass,
and threshold for Fig. 1, two stakes and two pieces of rough
board for potting-bench *m*; 70 feet of scantling, 2½ by 3
inches for spars to roof of Fig. 3; a few feet of pantile lath
for ditto, and 100 pantiles; three well-glazed 2-inch lights
for Fig. 2, which can be bought ready-made and seasoned of
any of the hothouse builders, these being the only parts,
excepting the door for Fig. 1, that require a first-rate joiner
to execute; 100 feet box of glass of the exact size required
can also be obtained without difficulty, and will leave
plenty in hand for repairs. Anti-corrison paint, the best for
out-door work, with directions for using, can also be bought
with the glass, as well as a stone of putty, or the latter can
be made by any labourer, but is better if made some time
before using. A window of some kind, which will serve for
lighting Figs. 3 and 4, must be provided.

In constructing the back wall remember to turn an arch
where the boiler is to be fixed, to prevent the necessity of
weakening the structure by cutting away; also, to see that
at least one of the hot-water pipes has a saddle cast upon it,
for supplying moisture to the atmosphere. *f*, *p*, *q*, and the
dotted lines, Fig. 4, indicate the position for the boiler,
furnace, and ash-pit under the building.

It will be convenient, in this place, to say a few words
about the multiplication of plants from cuttings, for by that
method nine-tenths of all the bedding plants grown are propagated. By far the greater proportion of plants that are
multiplied by cuttings require artificial heat. Nevertheless,
cuttings of many tender plants may be struck in the open
ground, or in pots and in frames, without heat, during summer,
and in every case the mode of procedure is nearly the same.
Very much of what we have to say will be applicable to
summer propagation without artificial heat, though our business is more directly with the propagation of plants in spring
by means of the heat of a tank or a hot-bed, because that
system must be resorted to with many bedding plants, and
requires more care than propagating in the open ground
during summer. We must suppose the heat to be sufficient and
constant. If from fermenting material, there should be a
large body of it in a nicely-tempered state. There is nothing

so good as a tank, for the operator has thus complete command over his work, and can enjoy the comfort of a warm house while attending to his duties. As a rule, a bottom-heat of 60° to 70° will suffice for all kinds of bedding plants that are struck from cuttings. A temperature of 80° to 90° may be used by persons who have had much experience, but 70° should be the maximum for beginners. In a subsequent chapter the raising of plants from seeds will be treated in detail; and for that reason this method of procedure is in this section only referred to casually.

Plants to be propagated from in spring should be in a free-growing state, because the best cuttings are those of shoots newly formed, and the worst those from shoots of last year. If therefore the plants are not freely growing, the propagator must wait for them; and to promote free growth, the temperature of the house should be kept at from 60° to 70°, with a moderate amount of atmospheric moisture, and as much light as possible, so that the young shoots will be of a healthy green, and with short joints. Suppose we look over a lot of fuchsias that have been some time in a warm house, we shall find them full of little stubby side-shoots all ready

CUTTING OF FUCHSIA.

to hand, without demanding any particular skill to remove them. Select one of these plump shoots, of an inch or an inch and a half long, press the thumb against it, and it will snap away " with a heel "—that is, with a thickened base, the separation taking place at the point where it issues out of the old wood. When you have removed it, it will probably have such an appearance as in the subjoined figure. All that this requires for its preparation is to remove the bud which has just started near the base of the cutting, so as to leave a sufficient length of clear stem to insert the cutting in sand firmly. When so inserted, and kept moist, warm, and shaded, roots will soon be formed at the base; and as soon as the roots have begun to run in search of nourishment, the top of the shoot will begin to grow, which is the sign for potting off. But suppose we have a chrysanthemum instead of a fuchsia. This will have a mass of tender shoots rising from the root, and

there is no need to take any of these off with a heel. With a knife, a pair of scissors, or the thumb-nail, remove a small shoot of not more than three or four inches in length —two inches will be sufficient. This will probably have some such aspect as in the figure. All the preparation this requires is the removal of the lower leaf, to make a sufficient length of clear stem for inserting it in silver sand. Or suppose we have instead a hard-wooded plant of robust growth, and which is known to be easily rooted, then we may venture to take a still larger cutting. The figure on p. 53 is a side-shoot of Veronica Lindleyana ; it consists of four joints, is young, the wood not yet hardened, and needs no preparation at all, because there is a proper length of stem for its insertion. In the case of plants having large fleshy leaves, it may sometimes be needful to crop off half of every leaf except those next the top bud ; but, as a rule, as many leaves should be allowed to remain as possible, because the more leaves that can be kept alive while the cutting is making roots, the quicker will it become a plant. No definite rule can be given on this head to guide the inexperienced. It all depends upon how many leaves can be kept alive. If the cuttings are to enjoy a brisk heat, say 70°, with plenty of atmospheric moisture, then nearly

CUTTING OF CHRYSANTHEMUM.

all the leaves may be left entire, and especially if the cuttings are in a close propagating frame, or under bell-glasses. But if they are likely to be exposed to draughts, if they are placed in pots or pans in an ordinary greenhouse, and, therefore, subjected to evaporation, the leaves must be reduced in number, and all the larger ones must be cut half away.

Another matter of importance in making cuttings is to determine whether they are to be rooted from a joint or not. Most cultivators prefer to cut the shoot quite close under a joint, so as to obtain roots from that joint. But there is no occasion to cut to a joint; any plant ordinarily propagated for the garden, will root as quickly from the "internode"—that is, the portion of stem intermediate between two joints—as from the joints themselves. This is of great importance when cuttings are scarce : as a shoot will often furnish half-a-dozen cuttings, if taking them at a joint is of no consequence ; and only one or two, perhaps, if taking them at a joint is imperative.

The size of the cuttings is a matter of great importance. As a rule, the smaller they are the better. Still, if very soft, many may damp off unless very skilfully handled, so the amateur must secure them moderately firm. 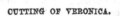 CUTTING OF VERONICA.

Three or four joints will generally suffice of most things, or say nice plump shoots of from one to two inches long. If young side-shoots are scarce, longer shoots may be cut up in lengths of three joints ; and if it is a question of raising the largest possible number of plants from the fewest cuttings, then one joint and its

accompanying leaf will suffice. Suppose we have a shoot
of a verbena placed in our hands to make the most of it ; we
shall first cut it into as many lengths are there are joints,
leaving each leaf untouched, and to every joint as much stem
as can be got by cutting just *over* instead of just *under* the
joints. Then with a sharp knife we split each of these joints
in half, so as to have one bud and leaf to each split portion,
and from every one of these we expect a good plant.

The most convenient way of disposing of the cuttings is
to dibble them into shallow pans filled with wet silver-sand,
as fast as they are prepared. The best way for those who
may have to leave the cuttings in the pans for any time after
they have formed roots, is to prepare the pans with crocks
for drainage, and over the crocks to spread an inch of chopped
moss or peat torn up into small shreds, or cocoa-nut fibre
dust, and then fill up to the brim with clean silver-sand. The
sand should be quite wet when the cuttings are inserted ; and
when they have been regularly dibbled in with the aid of a
bit of stick, or with the fingers only, it should be placed
where there is a bottom-heat of 60° to 70°. A temperature of
80° is allowable when time is an object, but at 60° better
plants may be grown ; in fact, there is generally too much
heat used. From the time of putting the cuttings in heat
till they begin to grow, the temperature must be steady, and
there must be regular supplies of water. But water given
carelessly will surely entail losses. Probably the sand will
retain sufficient moisture for eight or ten days, without needing to be wetted beyond what reaches it in the process of
dewing the leaves. To dew the leaves neatly and timely is
one of the most important matters. For the amateur, to
whom a few minutes is no object, the best way is to dip a
hard brush in water, then hold the brush beside the cuttings,
and draw the hand briskly over it. This causes a fine spray
to be deposited on the leaves, to prevent flagging ; but if the
water is given from the rose of a watering-pot, the cuttings,
if small, may be washed out of their places, or may be made
too wet.

A valuable contrivance for propagating plants in a sitting-
room, or in a greenhouse, is the Propagating Case of Messrs.
Barr and Sugden, of which figures are subjoined. It consists
of a frame containing a bed of moist sand, on which to place
the pots, and a boiler beneath, which is heated by a colza

lamp or jet of gas. It is an elegant hotbed in miniature, and will be found as entertaining as it is useful in the multiplication of bedding plants.

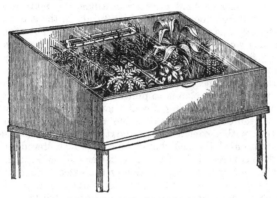

PORTABLE PROPAGATING FRAME.

Less difficult, but of far greater importance to the amateur, are the methods adopted for propagating plants during the summer and autumn. A few amongst the bedding plants may be multiplied by simply dividing the roots and planting again in a shady spot, or potting the divided pieces in small pots in which they are to be wintered for planting out in the parterre the following spring. But a majority of the most useful plants are multiplied by cuttings during July, August, and September, and are thus well rooted for storing in pits and frames during winter. In every case that admits of this practice it should be adopted, both because it occasions the least amount of labour, and insures far better plants than can be obtained by propagating in spring. The most important of the subjects requiring to be propagated in summer and autumn are geraniums and calceolarias. The first of these may be planted in any open border in the full sun, but it is better to prepare a somewhat sandy plot of ground in a partially shaded spot, for although the fiercest heat of the sun will not kill geranium cuttings, it is not altogether beneficial. Prepare the cuttings from ripe stout shoots, rather than from the softest green shoots, though if you begin in July, the softest shoots may be rooted if favoured with a little extra care. Cut them into pieces averaging three or four inches in length, and remove only as many leaves as will make half the length of the stem bare for inserting in the ground, remembering always that the more leaves that can be kept alive on cuttings, the more quickly will the cuttings root. Insert them firmly in the soil, and as close together as possible, so that their leaves do not over-lap. While dry weather continues sprinkle them with water every evening for a fortnight; and thenceforward, until they are taken up, do not give them a drop. In the cool autumn the calceolarias should be propagated, but in a somewhat different manner. The best plan is to prepare for the cuttings beds of light rich earth, consisting, for the most part, of leaf-soil, or peat, or thoroughly decayed stable manure in a pulverized condition, with a considerable admixture of loam and sand. The situation of the bed should be dry and sheltered, and it should be covered with a frame; or, better still, make up a bed in a brick pit, with a view to leaving the calceolaria cuttings where they are first planted for the winter, as there is no necessity for potting them if they can be protected from damp and frost, as they are almost hardy. In great establishments, where hundreds of thousands of bedding plants of all kinds are used, the greater part are struck and wintered in pits which rise only one to three or four feet above the ground level, and have the aid of hot-water pipes sufficient to keep the contents safe from frost. As a matter of course, whatever cuttings of tender plants are struck in the open borders must be taken up and potted for preservation through the winter, and this should be done as early as possible after they have made good roots, and before they begin to acquire a luxuriant growth in consequence of the warm autumnal rains. The small earthenware boxes and frames known as "Rendle's" and "Looker's," which have been described and eulogized in all the horticultural periodicals, are worthy of all the praise that has been bestowed upon them, for they can be employed to assist in summer propagating, and for preserving nearly hardy subjects through the winter, and for raising seeds on sunny borders in spring, and for many other purposes of the utmost importance to the amateur, and especially such an one as cannot boast of capacious plant-houses and endless appliances in aid of cultivation.

The great demand for space in greenhouses and pits is a matter of consideration in every garden where bedding plants are cultivated. The practitioner will not proceed far without discovering that it is a matter of the greatest importance to make the most of the daylight that can be obtained in combination with shelter. Various contrivances are resorted to with this object, and of necessity rectangular receptacles of some sort or other take precedence of circular ones in the economy of space. Bedding plants thrive in a most satisfactory manner in all their earlier stages of growth in shallow wooden boxes, and in many cases these may be obtained from the household store free of cost. In any case the boxes should be comparatively small for convenience of lifting, but the only important point is that they should be shallow, say, averaging four inches in depth, or six inches at the utmost. Having command of waste boxes and waste cardboards in considerable quantity years ago, we adopted a mode of combining the two which resulted in a great saving of both space and labour in the propagation and preserving of bedding plants. We must endeavour to explain the method, because to many a reader it may prove invaluable.

In the diagrams on p. 58, the first represents the box ready for use. Each compartment is filled with suitable compost, say, loam two parts, leaf-mould one part, sharp sand one part. The little seedlings or newly-struck cuttings are planted in the divisions singly, and at planting-out time each plant is presented to the hand in a single square block; there is no division necessary, not a fibre as fine as gossamer need be injured or disturbed. The sides and bottom of the box are wood; the divisions are thick cardboard. Suppose a fig box with the bottom knocked out. Now, across the bottom, at each end, nail a strip of wood. Next cut a piece of thin wood to make a loose bottom, the full size of the box, and drop it into the box to rest upon the two slips. Suppose the cardboard divisions next inserted, then, by turning the box on one side, and placing both hands against the loose bottom, as in Fig. 3, a little pressure with the fingers would thrust out the loose bottom and the cardboard divisions. The two slips over which the hands pass remain firm, because nailed down to the bottom edge of the box. You have only to suppose the divisions filled with plants, and Fig. 3 would explain the process of "turning out" not one from a pot, but fifteen from a box. The bottom being loose, yields to the pressure of the hands, just as the large crock in the bottom of a pot yields to the pressure of a

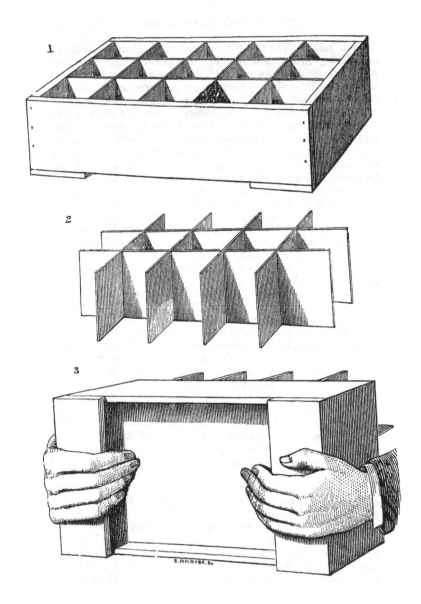

finger if the pot is inverted. But the contrivance is not used
in such a way at all. It is so engraved in order to convey an
accurate idea of its construction. When full of plants it has
but to be lifted on to a brick and the surrounding sides drop
down and leave the soil divided by the pasteboard in the most
handy position possible for operations. These pasteboards
are all that remain to be explained. They are first cut to fit
the box, and then are slit half-way so as to fit together firmly,
the short cross pieces being slit from the side which forms
their bottom edge, and the long pieces from the side which
forms their top edge. As they fit together firmly, each division remains intact to the last. Then, to liberate each block
for planting, the cross pieces are successively removed, which
frees the outside blocks, and, lastly, the two long slips are removed and the remainder are ready. Those who suppose this
to be a frail affair are mistaken. The cardboard will last two
seasons, and the wood-work a life-time. Any sized box that
can be lifted easily when full of soil can be employed in this
system of plant culture, but as it is well to name a size, it
may be understood that they should be eighteen inches long,
ten inches wide, and five inches deep. This will allow of
divisions three inches square, in which a very large amount of
soil may be placed. Those who can obtain waste card, which
is largely produced in some businesses, may grow all their bedders in this fashion. Probably zinc would be equally manage-
able, but not having tried it, we name it only at a rude guess.

Having got into winter work we present here figures of a
frame suitable for keeping calceolarias, the silvery-leaved cine-
rarias, and the centaureas,
if placed in a dry and sheltered position; with pro-
tective cover to keep out
frost, and a rack within
which the mats, or frigi-
domo, used during win-
ter may be stored away
neatly, and covered with
the glass lights placed
sloping upon it. ·The protective frame is made the
size of a frame-light, with diagonal braces, and bound at the
corners with iron hoop; and at each end is a small chain with

T link, to drop into a staple fixed in the frame or pit, by
which means the frames are secured in their places. The

scantling of timber used
is 2 inches by 1½ inch,
upon which is strained
stout canvas, projecting a
little over one edge of the
frame so that, when more
than one is required, the
projecting edge laps over
the next light, and keeps
the wet from going be-
tween.   After it is strained upon the frame, it should be
well painted—the frame should have been painted before.
The rack, upon which the lights are stored when not in use,
may be made to hold any number required. The timber used
for the rack must, of course, be of much larger scantling than
that for the frames.

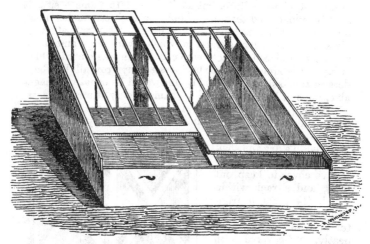

Preparations for planting out must be made early in the
spring. Plants that are tall and gawky must be cut back.
Plants that are in a crowded, starving state in pots or boxes
should be shaken out and potted separately. Plants from

which cuttings are to be taken should be put into a warm place, to promote a free growth of young shoots for the purpose. Plants that are comparatively hardy should be taken from the greenhouse and other warm places to cool pits and frames, where they may be inured to the air by degrees. At last, when the weather is favourable, about the end of April, or a week later, a regular movement must be made for a general " hardening off " of the entire stock. Begin with calceolarias and other nearly hardy plants. Follow with the hardiest of the geraniums ; and, as the season advances, proceed until the tenderest plants, such as coleus and alternanthera, have been turned out. But, where are they to go? Common pits and frames of the roughest sort will answer the purpose, and even old boxes on which boards can be laid may be turned to account, because while the sunshine and the air is mild, the plants are to be fully exposed; but are, at first, to be covered up at night, and during great part of the day, if the weather is unkind. Here is a figure of a " cradle "

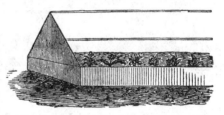

for hardening plants, and as it is the best contrivance of the kind in use, it will be proper to explain its construction. The cradle is four feet wide, and of any length required. The sides and ends are formed of deal planks, nine inches wide. These need not be planed, but a little neat carpentering will render them more durable, as well as more sightly. A bar of scantling forms the top, or ridge, and a similar bar is fixed on each side to form the slope. The bed is covered with clean gravel or coal-ashes, and on it the pots are placed. For covering, mats or cheap canvas may be employed ; and it is a small task, night and morning, to cover and uncover, and give a little water to the plants if they require it. As a rule, however, one part of the hardening process consists in keeping the plants rather dry, that they may the better endure the change of temperature consequent on their first experience of out-door life. A cheaper cradle may be extemporized by means of posts and nets, as shown in the figure on the next page.

It is the common belief that bedding plants require no
preparation of soil for their well-doing out of doors; and to
that belief we may attribute a large proportion of the failures
that occur, especially in unfavourable seasons. In peculiarly
kind summers, when showers and sunshine alternate, and
no extremes of heat or cold occur, the shortcomings of
the cultivator may not be made manifest; but in such a
summer as 1860, when there was no sunshine, but continuous
rain instead; or in such summers as 1868 and 1870, when
there was not a drop of rain for months together, but tropical
heat instead—then it is that real cultivation is plainly distinguished by its results from the slipshod pretence of cultivation that begins with a hoe and ends with a rake, and knows
nothing of the soil at one foot depth from the surface. The
fact is, flower-beds need frequent deep stirring and periodical

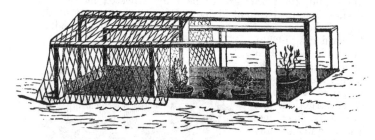

manuring, and the several beds, in a scheme for which many
sorts of plants have been prepared, should be severally prepared to receive them. This direction imparts an air of com-
plication to the business, but we cannot twist nature to suit
the indolent gardener. We must keep the truth in view, and
advise those who cannot grow bedding plants to do without them
altogether; for a shabby geometric garden is one of the shab-
biest of shabby things to be found amongst the demonstrations
of pretentious gentility. But the special preparation of every
bed for the plants it is to receive is not so serious an affair as
it looks. The fact is, the best general preparation is a deep
stirring of the soil at every change of crop, and the incorpo-
ration—a week or two before planting, if possible; but, if
not, immediately before planting—of a sufficient dressing of
manure for the plants that require it. Free-growing plants

of robust habit, such as geraniums, do not require manure,
unless the staple soil is exceptionally poor. Succulent plants,
such as echeverias and sempervivums, will do better in a poor
soil than a rich one, and may be aided by the addition of
sand and broken chalk or plaster to the plots they are to fill.
A few peculiar-habited plants, such as calceolarias, verbenas,
and lobelias, really require a rich and very mellow soil, and
the beds they are to be planted in should be dressed with
well-rotted hot-bed manure and leaf-mould some time in
advance of planting.

The distances at which the plants are placed in the beds
must be regulated both by consideration of their habits and
the requirements of the cultivator. In town gardens that
are required to be very gay during June and July, and may
be anyhow in the later months of the summer, the bedding
plants should be crowded in the beds to produce an effect
quickly. In gardens that are not likely to be seen until
August and September, thin planting may be practised, and
all flower-buds must be picked off as fast as they appear
until about six weeks before the "opening-day." In the
generality of gardens, the plants should be put so close that
they may be expected to meet by the middle of July. This
method will allow of the taking cuttings at the end of July
without serious damage to the beds; but it is always preferable to have a reserve of all the more important sorts in the
reserve ground expressly to cut from, so as to avoid even a
temporary diminution of the splendour of the parterre in the
very height of the season. It is the custom in some gardens
to take cuttings, in a wholesale manner and all at once, in the
early part of August; and the result is, the beds for a fortnight afterwards look as if they had been mown for hay, and
for the next fortnight they look so green and flowerless that
they ought to be mown again. It is astonishing how many
absurdities belong as it were of necessity to the bedding system, though of necessity they are all extraneous to it. As for
taking up and storing before winter returns, only one remark
shall be made. Take up in good time, and pot and house
with care, whatever is worth keeping and is really wanted.
But make no scruple of destroying whatever is not worth
keeping, or is not wanted, and let the destruction be accomplished in a quick and cleanly way. Our way is, to pull the
plants out and lay them in a heap, then to remove the top

soil of the bed, and throw in the heap of plants, and dig
them into the second spit or undercrust, and then to return
the top soil, and at once plant bulbs for spring flowering.
This is better, we think, than storing up the plants in a heap
to poison the atmosphere, or than allowing them to remain in
the beds after their beauty is past—a bugbear to the eye, and
a testimony of slovenliness.   Tastes differ, and "every one to
his taste" is a motto that very well suits a land of liberty
and a nation of gardeners.

CHAPTER V.

A SELECTION OF BEDDING PLANTS.

THE preceding chapter has disposed of all the general matters in connection with the cultivation of bedding plants, and it now becomes necessary to enumerate the principal subjects employed in bedding, and offer a few practical observations on their characters, uses, and the most convenient modes of cultivating them. The alphabetical arrangement adopted will facilitate reference, and it is hoped the very brief hints on selecting and multiplying the several varieties will be found sufficient, in connection with advices offered elsewhere in these pages.

AGERATUM.—This is one of the most popular plants, and is especially useful for ribbon borders, for the contrast of its grey or bright blue flowers with scarlet geraniums or crimson petunias. All the varieties may be raised from cuttings in spring; but it is far better to raise from seed, as they occasion less trouble, and, as a rule, maintain their character sufficiently for practical purposes. *A. Mexicanum*, the species originally cultivated, has of late been greatly improved in a series of dwarf varieties. The dwarfest of them is *Tom Thumb*, which, in ordinary soils and situations, does not exceed six inches in height. It is most valuable for edging purposes where a band of pale blue is required next grass or gravel, and for panel beds, because of its neat habit. It, however, is hardly robust enough for very poor soils. *Imperial Dwarf* usually attains a height ranging from six to nine inches, and is remarkably neat and compact in habit. It can be employed as a first or second row plant, or as a centre in small beds. *Prince Alfred* is rather taller than either of the preceding, and is therefore more valuable for large beds or ribbon borders, excepting for outside rows. The seed should be sown in a gentle heat in February, and when the plants are well established in the small pots in which they

5

are put from the seed pots, they must be stopped, to induce them to produce side-shoots and form bushy specimens by planting-time.

ALTERNANTHERA.—This genus presents us a series of lovely little plants with red, crimson, or orange-tinted leaves, which are valuable for edging flower-beds, and eminently so for leaf-embroidery. They are tender plants, and require more than ordinary care in their cultivation; hence it is not unusual to hear them spoken of slightingly as comparatively worthless. As to planting them, it is utterly useless to put out plants that are not much larger than "darning needles;" for, if they grow at all, the summer will be past before they can fill the space allotted to them. To insure bushy little plants by the end of the spring, the cuttings should be struck moderately early in the autumn, and wintered near the glass in a stove or warm greenhouse; during the early spring months, after the plants are potted off separately, they can be grown in a comparatively cool temperature. It is not necessary to pot them off in the autumn, and a few panfuls of cuttings take up but little space. When the propagation of the bulk of the stock is deferred until spring, it is difficult to strike the cuttings early enough to afford the plants sufficient time to attain their proper size by the planting-out season; to force them in much heat will render them less able to battle with unfavourable weather when first put out in their summer quarters. It is also important to plant them in beds situated in a warm sheltered situation, and they do much better if the surface of the bed is elevated a few inches above the general level. The best of the varieties is *A. magnifica*. It combines the vigour of growth of *A. paronychioides* with the rich leaf-colouring of *A. amœna*, both of which are good. *A. spathulata* and *A. versicolor* are of little use for outdoor work.

AMARANTHUS.—The most popular plant in this section is *A. melancholicus*, with beautiful claret-coloured leaves. The scarcely-known *A. tricolor* presents the most brilliant leaf colours of any plant in the garden, but is peculiarly tender and is rarely seen well grown. A newer variety, called *A. elegantissimus*, equals tricolor in beauty, and is somewhat more vigorous in habit and more constant in colouring. They are all raised from seed, which should be sown in March in a gentle heat, and the seedlings pricked out into pans and boxes as soon as they are large enough to handle. They may be sown in April and May, for planting out after early-flowering annuals, and for succeeding other subjects that do not last the season through. It is of the utmost importance to obtain good seed and to sow plenty of it, for in spite of the greatest care in saving the seed a large proportion of the plants of the two last-named kinds will be so poorly coloured as to be useless for bedding purposes; but when the seed is saved carelessly not more than twenty-five per cent. of the produce will be of any use. When they are strong enough for potting off, destroy all that are entirely green, and pot the others off separately. The exact number of plants required should be determined upon, and thirty per cent. more potted than will be wanted, to allow of that number of the worst being discarded at the planting-time. By carrying out this plan, the plants with the most richly-coloured leafage only need be planted; and as no other plants suitable for bedding purposes possess such gorgeous leaf-tints, the appearance of a well-filled bed is most magnificent. *A. melancholicus* always comes true from seed, and is one of the easiest plants to grow that was ever seen in a garden.

BEET is not to be considered a desirable plant for the flower garden, but as it is often used, it must have a place here. The best for the flower garden are *Dell's Crimson* and *Barr's New Crimson Leaf*. Both are compact in growth and rich in colour. The roots are of fair average quality, and can be used for salads, in the same manner as the varieties usually grown in the kitchen garden. The seed can be sown in the beds early in May, or it may be sown in a cold frame in April and transplanted into the beds when the other bedders are put out. Beds filled with beet should have a broad edging of centaureas or variegated geraniums, to prevent anything but a level surface of leaves of the beet being seen. The soil should be dug rather deeply to enable the roots to strike down, or the roots will be forked and of no use for culinary purposes. The *Chilian* beet, excepting for wildernesses, is worthless, and should not be grown.

BOUVARDIA.—A small bed of mixed sorts has a very pretty effect, especially if a few plants of the blue-flowered Plumbago capensis are mixed with them. The bed in which the Bouvardias are to be planted should be prepared, if practicable, as advised for the Lantanas, and strong bushy plants should be put out. They are usually propagated in July, and potted off when nicely rooted into three-inch pots. Then by the middle of March they are strong enough to be shifted into pots one size larger. We stop them about three times during the period between their being potted off singly and their being put out in the flower garden, and by the end of May they are usually bushy little specimens and begin to flower at once. In the autumn a dozen or so are taken up and carefully potted and placed in a warm greenhouse, where they bloom profusely throughout the winter. The best for the flower garden is *B. angustifolia*, which is of no use for the winter. *Brilliant*, cerise; *Delicata*, pink; *Elegant*, scarlet, and *Hogarth*, scarlet, are also good.

CALCEOLARIA.—The chief fault of the calceolaria is its inconstancy. It is not uncommon for all the calceolarias in the country to perish about the middle of July, leaving the parterres they should have adorned with masses of golden flowers abominably ugly with their withered stumps, or, at the best, obnoxious blanks. In the experimental garden at Stoke Newington the cultivation of this plant has received considerable attention, and it is believed that every difficulty experienced by amateurs may be overcome by the adoption of the system of cultivation which will now be recommended. The only varieties suitable for bedding are those of decidedly shrubby habit, which produce comparatively small flowers. Those that have somewhat soft stems, and large leaves and large flowers, partake too much of the characters of the herbaceous section to be fit for battling with the vicissitudes of outdoor life, and, moreover, they always produce their flowers in a series of efforts, and not continuously. The proper time to propagate them is from the middle of September to the middle of November, when they do not require heat; but they may be very quickly multiplied by cuttings in a gentle heat in spring; and if the summer is favourable to calceolarias, spring-struck plants do well, though they do not begin to flower so early as those struck in autumn. There can be no better method of procedure than to make up a bed of light soil, consisting of such materials as leaf-mould, sweepings of a manure heap, half-decayed moss, and the sandy stuff thrown out of pots in the potting-sheds. The bed should be in a frame or pit, within a foot of the glass, or on the border of a cool vinery or peach-house, as near the glass as possible. Prepare the cuttings from soft side-shoots, and plant them

firmly in the bed, about three inches apart, and keep them
regularly sprinkled to maintain their freshness until they
are rooted, after which they will require but little more attention beyond watering, ventilating, and protecting from frost.
They must be wintered rather dry, and have plenty of air, or
many will perish. There is yet another extremely simple, but
most effectual, method of procedure. Its first requisite is a
greenhouse, or pit, which is sufficiently heated to keep out
frost. In this structure the bed is to be made up near the
glass, of some such light kindly soil as recommended for the
frame. Leave the plants in the ground until the middle of
October, or, if the weather permits, until the middle of November; then pull them to pieces so as to separate the best
young shoots with a heel, and strip the bottom leaves from
them, and dib them in, and press them firm, and the work
may be considered finished. As for the roots, throw them
away. They may be crowded together so as to make a solid
field of leafage; but, as a rule may be useful, we will say plant
them three inches apart. A slight sprinkle over the tops
occasionally will be good for them, but they must be kept
rather dry, and must have a little heat to help them through
frosty weather. No matter which of these two methods be
adopted, the whole of the plants must be lifted in the first
week of March, and be planted out in beds of light rich earth
in frames facing the south, where they will make rapid progress if taken care of. The latter part of the month of April
is the proper time to plant calceolarias; if the planting is
deferred the plants are endangered. The beds for calceolarias
should be prepared by deep digging and liberal manuring with
rotten hotbed manure and leaf-mould; and if there is no leaf-
mould at command, use an additional dressing of the hotbed
manure in place of it. If planted in poor ill-dressed soil, the
plants are endangered; in fact, the principal reason of the
failure of the calceolaria in a hot dry season is defective root-
hold, the result of planting late in poor soil, the plants having
been already nearly starved to death in pots as a preparation
for their final extinction. In the operation of planting the
plants should always be put into the ground as deep as possible, but of course without burying the branching portion of
the stem. Deep planting encourages the formation of a fresh
set of roots, and places the roots already formed at the greatest
possible distance from the surface, where they are compara-

tively safe against the exhaustive action of a hot sun. As to
watering, one or two liberal doses may be given within the
first ten days after planting, but it is far better to give none
at all if only the ground is moist enough to carry them on
safely until the next rains occur. A considerable quantity of
bedding plants are killed every year by watering them, or
rather, by tormenting them, with a pretence of watering. As
for varieties, there are not many good ones, but the few that
are most worthy of attention are wondrously brilliant if they
happen to behave well. *Amplexicaulis* is the tallest in growth,
the flowers are palest yellow. *Aurea floribunda, Canariensis,
Gaines's Yellow,* and *Golden Gem* have deep yellow flowers,
and in habit are dwarf and compact. The red and brown
varieties are simply useless.

CENTAUREA.—The silvery-leaved centaureas are among the
most striking and valuable of the leaf plants we possess, and
especially to contrast with the crimson and chocolate-coloured
coleus. Fortunately they are very hardy and quite easy to
grow, though there has been much said about the cultivation
of them by folks who sought or imagined difficulties. The
best way to raise a stock is by sowing seed. Many cultivators find it an easy matter to strike cuttings in autumn, and
others cannot accomplish the feat. But any one may strike
them in the spring and insure a stock with ease and rapidity.
Before spring-cuttings can be obtained, however, a sufficient
number of old plants have to be housed and carefully attended
to throughout the winter, or a large proportion will rot off
just above the soil, and, as a matter of course, perish. Having
brought them safely through the winter, we have to place
them in a genial temperature of between 55° and 60° soon
after Christmas to start them into a steady growth. If all
goes on right, they will produce a nice crop of cuttings,
which, if taken off with the smallest heel possible, inserted in
cutting pots, and the pots plunged into a brisk bottom-heat
in the propagating pit, a large proportion will soon strike and
in time make plants. This manner of dealing with them in
large gardens is by no means difficult, as there will be a
peach-house or vinery at work in which the old stock plants
can be placed, and also a cucumber bed or a propagating pit
in which the cutting pots can be plunged. But what can the
owner of a greenhouse and a few pits do with them? Simply
nothing!

This consideration brings us back to the subject of seed, and it is by no means bad practice for the amateur to save his own. A stock of old plants must be kept for this purpose, and for small gardens about twelve or fifteen will produce sufficient. That number should be put by at planting-out time every year, and be potted into six-inch pots, and the pots plunged in a bed of coal-ashes in the full sun. They should remain in the open until the end of September, and then be placed in a cold frame, where they can have protection from heavy rains, and, at the same time, be exposed to the air. In the spring and early part of the following summer they will flower profusely, and, if placed in a light, airy position, will produce an abundance of seed. The seed must be sown as soon as gathered, in pans in the usual way, and the pans must be placed in a cold frame. The seedlings, if potted off when of a fair size into small pots and kept close for a fortnight, will make nice plants for the following season. The young stock should be removed to the open when nicely established, and be wintered in a cold frame, with a mat or litter thrown over the glass to keep out the frost. The greatest enemy to centaureas during the winter is damp; therefore no more water must be administered than is really necessary.

The two best for front lines are *C. argentea plumosa* and *C. ragusina compacta*. Both are very neat in growth, and wonderfully effective. *C. gymnocarpa* is immensely valuable for back rows and centres of large beds, and *C. ragusina* is equally valuable for second rows. Both grow rather coarse in rich soil, and therefore when employed in conjunction with subjects that require a liberal share of nourishment, they should be plunged in the beds instead of being turned out of the pots.

CINERARIA.—The "silver-frosted" plant has lost much of its popularity within the past few years on account of the superior claims of the centaureas. But it is, and always will be, valued for its distinct silvery-grey colour, neat habit, and hardiness. In all respects the cultivation should be the same as advised for the centaureas. *C. maritima* is the best known of the series, but *C. acanthifolia* surpasses it in silvery whiteness and is to be preferred.

CLEMATIS.—The new garden varieties may be described as "sensation" bedders, for if large compartments can be devoted to them, they produce a wonderful display of crimson, purple, and blue flowers. They will thrive in any good garden soil that is well drained, and, generally speaking, adapted to bedding plants. But the soil which suits them best is a light, rich, sandy loam ; the lighter the soil the better, but it cannot be too deep or too well drained. They are hardy enough for all except the bleakest climates in these islands, but a warm sheltered position and full exposure to sunshine are conditions that conduce greatly to their prosperity, and, consequently, to their abundant flowering. They should be planted two to three feet apart in large clumps. A number of varieties may be mixed, as they all present shades of crimson and purple, but the most decided effect will be produced by employing one showy variety for a bed, or a number of varieties distinctly arranged in bands or rows. Some time in June the beds should be covered with two inches depth of half-rotten manure, put on carefully. The plants will soon cover and hide it, and will enjoy the moist surface it will insure them during the heat of the summer. As the plants progress they must be pegged down a little higher than verbenas, and quite as regularly. All the growth they make should be left until the month of April following, when the whole of the plants should be cut back to within six inches of the ground. The best way to multiply them is to put down layers in August, but young shoots may be struck under hand-glasses in June. The best varieties for bedding are : *Jackmani*, violet purple ; *Rubro-violacea*, reddish violet ; *Rubella*, deep claret ; *Viticella amethystina*, pale violet blue ; *Tunbridgensis*, dark blue ; *Lanuginosa pallida*, lilac ; *Lanuginosa candida*, white.

COLEUS.—A few of the varieties of coleus are gorgeous in their leaf-colouring and invaluable as bedders, but some thirty or forty kinds, supposed to be " in cultivation," are scarcely better for outdoor purposes than nettles from the hedgerows. To grow these plants is easy enough, provided they can be wintered in a stove or intermediate house, and be propagated early over a tank or on a good hotbed. They cannot be wintered in the cool temperature that suffices for geraniums, centaureas, and verbenas, and it is but inviting vexation to attempt it. But given warmth enough and the matter is disposed of, for they grow with the vigour of nettles if they grow at all. During winter keep them rather dry and near the glass, and never allow a drop of water to touch the leaves. Early in spring strike the cuttings in a moist heat of 70°, and pot off the young plants in a light rich compost. Do not be in haste either to turn them out to harden or to plant them in the beds. The middle of May is early enough to put them in frames, and the first week in June early enough to plant out. The best of all the varieties is *C. Verschaffelti;* but *Emperor Napoleon, Princess of Wales,* and *Baroness Rothschild* are useful where more than one sort is required.

DIANTHUS, OR INDIAN PINK.—The varieties of *Dianthus Heddewegi* are so wonderfully showy, and so easily raised, that it is surprising they are not more generally grown. *D. H. diadematus fl.-pl.* and *D. H. laciniatus fl.-pl.* are, perhaps, the most valuable. A packet of seed of each of the varieties will yield a number of colours of the most attractive character. They are not so suitable for geometric schemes as many other things, and should be planted in beds that stand out singly upon the lawn. Sow seed in February or March, and if any of the plants produce peculiarly fine flowers, take cuttings of them in June or July, in order to keep them true.

ECHEVERIA.—The species of echeveria are valuable in eccentric bedding, and for edgings to leaf-embroidery. They may be propagated by seed, offsets, cuttings, and leaves. Plants raised from seed sown in spring will not attain a size large enough to be of service in the flower garden during the ensuing summer, but they will be valuable for the following seasons. Echeveria seed is so minute that, like calceolaria seed, it will perish if buried too deep. Previous to sowing, it is essential to make the surface perfectly level with a piece of board, or one part of the seed will be buried to a great depth, and the other will not be covered at all. The seed-pans should be placed in a brisk temperature. Offsets may be taken off and potted at any time, except the depth of winter. To propagate by leaves, take them off the plant by a snap with the thumb, so as to have the base complete. To fix them base downwards on the soil, drive through every leaf a wooden peg; this should be done in July and August. To winter these plants, pot them in very sandy compost, with plenty of drainage in the pots, and keep them as near the glass as possible, and allow them but a moderate supply of water.

FUCHSIA.—This most elegant of greenhouse plants is of small value for bedding, but is occasionally employed with good effect in beds that stand apart from groups, and there can be no finer subjects than large pot fuchsias for the terrace walk, and to form groups on the lawn on *fête* days. To grow the fuchsia is like turning the key in a lock: turn it the right way, and the lock responds; turn it the wrong way, and resistance forebodes failure. In the early months of the year fuchsias should grow fast, and the principal agents to promote growth are warmth and moisture. In a dry, much-ventilated house, fuchsias fade away as if blighted, no matter how good the soil may be in which they are grown. But keep them rather close in a temperature of 50° to 60° or 70° as the season advances, and aid with frequent syringing from the time they begin to grow in spring until they no longer need artificial heat, and they will grow freely, even in a bad soil. But they require good living, and there can be no better mixture for them than two parts of mellow turfy loam and one part each of rotten hotbed manure and good leaf-mould. At all seasons they need more moisture at the root than the generality of greenhouse plants, and even in winter should never be quite dry. To propagate them is an extremely easy matter, and the best time for the amateur is in February, when young shoots an inch or two long strike as by magic in a moist heat of about 60°. When planted out select a rather shady moist situation, and prepare the bed with a good dressing of leaf-mould and manure. The light kinds make the best display in beds, but the dark grow most freely. The following light flowers are the best for beds : *Guiding Star, Mrs. Marshall, Brilliantissima, Minnie Banks.* The following dark varieties are fine: *Splendour, William Tell, King of Stripes, Model.* The variegated-leaved varieties are extremely showy, especially *Cloth of Gold, Meteor, Golden Treasure,* and *Regalia.*

HELIOTROPIUM.—If the heliotropes lack colour, they make amends by their delicious odour. A few of the newer kinds, however, present us with fine dark blue or violet flowers. They may be raised from seed or cuttings with the greatest ease by the aid of heat in spring. When planted out, a poor soil suits them best, and when housed for the winter they must have but little water, and never experience the slightest touch of frost. The best are *Florence Nightingale, Surpasse Gauscoi, Etoile de Marseille,* and *Modele.*

IRESENE.—The most valuable of all the dark-leaved bedders sent out for many years past is *Iresene Lindeni.* It is neat and compact in growth, possesses a hardy constitution, and the colour of the foliage is the richest sanguineous red imaginable. For centres of beds and second and third rows it is all that can be desired. It is more effective than any other plant of a similar character, and it can be wintered in a cool greenhouse; indeed, it has made us independent of the coleus, and is therefore of immense value to amateurs.

LANTANAS.—These are not well understood, or they would be employed extensively in the summer decoration of the flower-garden. All the varieties are not suited for the open ground, but a few are marvellously showy. They, however, do not flower in all soils and situations alike, as they require a warmer soil than the generality of bedders; but they will flower profusely in naturally cold soils if the precaution is taken to elevate the bed in which they are planted six or eight inches above the general level. In naturally warm and dry soils the beds should of course be on the level, or in very hot weather the plants will be dried up unless they are watered frequently. The soil of the beds should not be too rich or too poor, and should, if practicable, be dressed with leaf-mould or vegetable refuse, instead of with decayed hot-bed or stable manure. The cultivation may be generally described as corresponding with that of the pelargonium, but it is best to propagate by cuttings in spring in a brisk moist heat, and to winter the old plants in a rather warmer house than pelargoniums require. Old plants make magnificent beds in isolated sunny spots in a good season. The following are fine varieties:—*Dom Calmet*, lilac pink and yellow, compact in growth, and most profuse in flowering; *Jean Bart*, yellow and bronze, very dwarf and free flowering, one of the very best; *Monsieur Escarpit*, deep rosy purple, very distinct; *Cauvin*, yellow and rosy. The best of the older sorts are— *Alba lutea grandiflora*, white and yellow; *Impératrice Eugénie*, rosy pink, very dwarf and free, the best for edging purposes; *Adolphe Hwass*, canary yellow; *Mons. Rougier*, yellow bordered with reddish scarlet; *Rœmpler*, crimson and orange; *Roi des Rouges*, scarlet and orange; and *Victoire*, pure white.

LOBELIA.—The smaller kinds are immensely popular, but few amateurs grow the stately and gorgeous varieties of what is called the "herbaceous section," nor, sumptuous as they are, have they any right to special notice here, for they are not bedding plants. Our little friend, *L. erinus*, is the centre of the group, from which are derived the bright blue, deep indigo, grey, and rosy-flowering varieties in request everywhere for marginal lines and edgings ; the very perfection of bedding plants, which any one may grow with but a trifling exhibition of skill and patience, with the aid of a glass structure of some kind or other, it scarcely matters how rough and simple. We have before now made a good edging of lobelias without taking a cutting or sowing a seed, for we have found self-sown plants in myriads, in pots of geraniums, and on beds of earth in the greenhouses, and even on brick walls and planks under glass everywhere ; and have left them alone until wanted, and then, on some mild, cloudy day, have transferred them to the open ground, and left them to settle accounts with the weather and take their chance for weal or woe. The best plan to adopt for securing a good stock is to raise the plants from cuttings, selecting for the purpose the best varieties obtainable. In the first place, plant out a few of the selected sorts, at the end of May, in some out-of-the-way place, and' let them grow and flower as they like. About the middle of July, cut them down pretty close to the ground, and they will soon after bristle with new tender shoots. These must not be allowed to flower, but, as soon as they have attained a length of about two inches, take them off, and dibble them into a bed of sandy soil, in a frame, or under hand-glasses or propagating-boxes, and keep them shaded and sprinkled until they have made roots. If they run up quickly to flower, nip out the flower-buds to keep them stocky and strong. Take up early, and pot carefully, and keep near the glass all winter, never allowing them to flower. In February these will supply cuttings in great quantity, and any one who can strike a cutting may make a good plant of every one of them. To raise them from seed is a still simpler matter, and if the seed has been carefully saved, the plants will be tolerably uniform in character, and will be good enough for large gardens, where a few spurious plants in a mass will not be noticed ; but seedlings are not to be depended on for highly-finished work. Sow the seed in pans or boxes of fine rich sandy soil, covering it with a mere dust of peat or finely-sifted leaf-mould. The seed need not be sown until March, as the plants grow rapidly when they have made a fair start. Lobelias should not be planted out in flower, or with the flower-buds visible. It is best to cut the tops off the plants a week before planting, which will promote a bushy growth, and prepare them to throw out roots vigorously when planted. If planted in flower, they may be expected soon after to go out of flower and remain blank for a month. If treated as here advised, they will be blank about a fortnight at first, and will then flower freely for the remainder of the season. None of the old varieties, such as *speciosa* and *gracilis*, are now worth growing, because better are at command. The very dwarf sorts, known as the *Pumila* section, are exquisitely beautiful, forming dense cushions solid with bloom of the most pure and brilliant colours. The most useful of them are—*Grandiflora*, deep blue; *Azurea*, light blue; and *Annie*, lilac. The following are also first-rate for various purposes in the parterre, and also make charming pot-plants :—*Indigo Blue*, intense deep indigo blue; *Spectabilis*, deep cobalt blue; *Trentham Blue*, clear blue, white eye; and *Mauve Queen*, rosy lilac.

MARIGOLDS are not to be despised, because the little orange-flowered *Tagetes* is one of the best bedding-plants known, and a capital substitute for the calceolaria on soils that do not suit that capricious plant. They are all grown from seed, and as to their requirements, they are real "poor man's plants." But let us consider the large-flowering marigolds first. The *Miniature* or *Pigmy* and the *Dwarf French Marigolds* must not be despised by those who have not the means of growing yellow calceolarias. They are very dwarf in growth, the varieties of the miniature section ranging from six to nine inches in height, and those of the dwarf section averaging twelve inches. They vary considerably in habit, unless unusual care is taken in saving the seed; hence it is most important to obtain it from a respectable source. The yellow-flowered varieties of both sections will be the most useful for bedding purposes. The brown and striped-flowered varieties are very distinct, but they can only be employed in the mixed border, or in an isolated bed. In ordering the seed, it will be necessary to state the colour required. The dwarf-growing *Tagetes signata pumila* is the most formidable opponent the yellow-flowering calceolarias have yet had to encounter, for it grows freely and blooms most profusely where the calceolaria cannot exist. Indeed, it ought not to be planted in very rich soil, because, when supplied with a large amount of nourishment, many of the plants will become over-luxuriant, and hide a considerable proportion of the flowers with the foliage. To raise a stock with the least amount of trouble possible, sow the seed in March, and place the seed-pans in a frame or greenhouse, and, as soon as the plants are well up, place the pans in a position where they will enjoy full exposure to the light, and a moderate amount of air. Prepare a bed of light loamy soil in a cold frame, or where it can be covered with lights for a few weeks in a sheltered corner, and then prick out the plants, as soon as they attain an inch in height, at a distance of not less than three inches apart. When the bed is filled, water liberally to settle the soil, keep close for a few days, and shade, to enable them to become established quickly. Afterwards ventilate freely, and when the weather will permit draw the lights off altogether. Coddling must be avoided in all stages, or they will be drawn up tall and lanky: a state of things by no means desirable. They should have no more protection, after they are put out in the frame, than is really necessary to protect them from sharp cutting winds and frosts. The seed-pans should not be placed in a propagating frame or other structure in which a high temperature is maintained. Plant out, nine inches apart, in May, and keep a watch over their growth. Any of them that threaten to make a rank growth should be destroyed, and the gaps will soon be hidden.

MESEMBRYANTHEMUM.—A few of these are invaluable for hot dry positions, and especially for sunny slopes and odd places, where ordinary bedding plants would be starved, or, if they prospered, would be too showy. The best rule for growing these is to strike cuttings in July, and winter them in sandy soil in a sunny greenhouse, keeping them rather dry. The best for bedding are *M. conspicuum, M. spectabile, M. formosum, M. blandum, M. glaucum, M. curviflorum, M. aurantium, M. lepidum, M. polyanthon, M. glomeratum, M. coccineum major, M. diversifolium, M. inclaudens, M. floribundum, M. aureum.*

NIEREMBERGIA.—One sweet little plant of this family is useful to make miniature masses and bands of comparatively unattractive white flowers, and especially useful to plant at the sides of rustic baskets, to fall over and make festoons and ringlets of fairy flowers. The stock for bedding purposes is raised from cuttings in spring; but old plants are best for rustic vases, and for clothing sloping banks. *N. gracilis* is the most useful; indeed, the pretty *N. frutescens* and *N. rivularis* are of no use as bedders.

PANSY.—A considerable number of fine bedding pansies have, within the past few years, been introduced to gardens as bedding plants, greatly to the disappointment of many who were not cognizant of their real characters, and who associated them with geraniums, verbenas, and petunias in the expectation of late summer and autumnal bloom. It may be said, with but small fear of contradiction, that no pansy or viola is adapted for the parterre in the later months of summer, except in a certain few localities; but many of them are invaluable anywhere and everywhere for their beautiful and abundant bloom throughout April and May; and therefore their proper association is with arabis, alyssum, and iberis, which flower long in advance of the summer bedders. The best of the bedding pansies should be kept true by growing them from cuttings, but they reproduce themselves tolerably true from seed, and this method of multiplying them is the easiest. Cuttings of pansies may be struck in a gentle heat in spring; but, to grow them successfully, cuttings should be planted on a shady border during the summer. The earlier they are put in, the more surely they form strong plants, and the earlier will they flower. The cuttings require little or no protection, except from sun, and to be kept sprinkled in dry weather. Plenty of young rooted pieces can also be taken away from the plants during the summer, and, if planted in nursery beds, they will make fine plants for removing into their winter quarters in October. Plant out in October if possible, *planting firm*, and in moderately good soil, not too rich, and the close-growing sorts closer together than the more spreading kinds. As pansies suffer far more from cold easterly winds than from any other cause, mulch the beds with either half-decomposed leaf-soil or cocoa-nut refuse. Through April, May, and June, shake a little sifted good soil about the plants and amongst the shoots, to encourage top roots; and, when the shoots are long enough to require pegging down, fix them neatly to the ground, to protect them from injury from wind, and induce them to throw out roots and side-shoots. The pansy does not require much water, but, in very dry hot weather, the beds should be frequently watered with a rose watering-pot, especially night and morning. Take especial care, in planting out in beds, to use young plants that were struck from summer cuttings or young offshoots. Old plants pulled to pieces frequently fail, or make only half the show that may be obtained from young ones. In the northern parts of these islands the climate is more favourable to pansies and violas, and they are more valuable as bedding plants than in the warmer south. But, as remarked above, for a charming display during April and May and some part of June, there is nothing more cheap and certain. No one in these realms has ever seen a bed of pansies covered with flowers between Midsummer-day and the first of August, and the autumnal bloom is never equal to that produced in spring. The following are the best varieties :—*Cloth of Gold*, yellow ; *Sunshine*, coppery orange ; *Imperial Blue*, light-blue purple ; *Dean's White*, white ; *Cliveden Yellow*, yellow ; *Magpie*, purple and blue.

PELARGONIUM.—Under this head we must consider the uses and characters of what are commonly called " bedding geraniums." We have here nothing to do with elaborate classifications, or with the various methods of cultivation by which exhibition plants and new varieties are produced ; therefore, though the subject might occupy a bulky volume, we shall hope to say enough for the present purpose in a contracted paragraph. The first thing to be said is that the zonate pelargonium is the king of bedding plants. It may be dispensed with, indeed, in particular styles of planting—as, for example, the sub-tropical—but there is no other plant capable of so many and such varied uses, and in some way or other it might be made to play a prominent part in almost any scheme of colouring that ever was devised. The wide range of its characters, and consequently of its uses, is in a wonderful degree enhanced in value by its hardy constitution, and the comparatively small amount of skill and labour required in its management. Sunshine it must have, and really that is about all it requires, if we may adopt a " rough and ready " mode of expression. Speaking of the family as a whole, it may be said that a somewhat poor soil suits them best, but, nevertheless, the beds should be well prepared for them, to encourage deep rooting early in the season, for a good root-hold is essential to long-continued flowering, especially in an exceptionally hot and dry season. Sandy and chalky soils should be improved for geraniums by the addition to the staple of thoroughly pulverized hotbed manure and leaf-mould, but all good loams of average depth are sufficiently nourishing and need not be manured. It is very bad practice

SHOW PELARGONIUM.
(French spotted.)

to water geraniums after they are planted out, but of necessity
if the weather is particularly hot and dry immediately after
planting, they must be assisted for a week or so. It is also
bad practice to put them out full of flowers; in fact, they
ought not to flower even while in the house, or in pits or
frames. But trusses will show themselves, and should be
pinched out before they open, and if they occur on shoots that
rise a little above the general contour, those shoots should be
at once cut back a few inches. The result of these precautions
will be to defer the first show of bloom in the beds somewhat, but when it appears it will be more solid and continuous than in the case of plants allowed to present odd
trusses in their own way from the month of February onwards.
All the varieties should be propagated in June, July, and
August, and be housed in good time, to prevent the rank
sappy growth that the warm autumnal rains are likely to
produce if they are left out too long. Winter them rather
dry, with abundance of light and air, and never give heat
beyond what is barely sufficient to keep out frost. Geraniums
and calceolarias will bear 5° of frost without harm in ordinary
cases; therefore the thermometer in the house or pit appropriated to these plants may sink to 27° safely. We have,
indeed, had thousands of seedling geraniums in a somewhat
sappy state through being grown from the first under glass
frozen to the extent of 10°, and have not lost a dozen in
consequence. But it is not well to expose plants to extreme
conditions, and the amateur cultivator is advised to maintain
the temperature of the geranium pit at all times a few degrees
above the freezing-point, for it costs little to be safe, and it
may cost much to go in the way of danger. It will not surprise the reader to be told that as geraniums differ in habit
and constitution, so they differ as to their requirements.
The differences, however, are slight, and may be disposed of
in a few words in connection with the several groups.

Single Red Zonals.—In this class we place all the scarlet,
pink, and purple varieties, whether they have broad or
narrow petals. In other words, we do not need a class for
nosegays. It is equally unimportant whether the leaves are
actually marked with zones or "horseshoes," or are wholly
green. The cultivation of these has been sufficiently described, and it remains, therefore, only to present a list of a
few of the very best for bedding. In selecting these, form is

6

of less consequence than colour, habit, and abundant flowering.
We begin, of course, with the pure Scarlet section, from which
we select *Thomas Moore, Orbiculata, Attraction, Bonfire, Cybister.*
From the Orange and Salmon tinted we take *Hibberd's Orange
Nosegay, Beaton's Indian Yellow, H. W. Longfellow, Harkaway.*
Crimson and Purple tinted: *Le Grand, Duchess of Sutherland,
Black Dwarf, Bavard, Waltham Seedling.* Rose and Pink
tinted: *Feast of Roses, Madlle. Nilsson.* Cerise: *Tristram
Shandy, Lion Heart, Lucius.* Lilac and Purplish Rose: *Lilac
Banner, Amy Hogg, Duchess, Lilac Rival.*

Single White Zonals.—All the white-flowering varieties
should be grown in a poor soil, and if the scheme of colouring will allow of it, a partially shaded spot will suit them
better than to be exposed to the full blaze of the meridian
sun. When grown in a poor soil, and enjoying morning and
afternoon, but not mid-day sun, the flowers are more pure
and more plentiful than in the case of rich soil and full
exposure. In the case of the beds selected for white geraniums
being too strong in texture and condition, plunge them in
their pots, and, if possible, get up a reserve stock of plants
to take their place if they should happen to become flowerless after July. The best are *White Wonder*, and *White
Princess.*

Double Zonals are not well adapted for bedding. Those
who are disposed to try them are advised to plant in a poor
soil. The best for the purpose are *Gloire de Nancy*, rosy
carmine; *Le Vesuve*, scarlet; *Princess Teck*, deep scarlet;
King of the Doubles, bright cerise.

Golden Zonals.—These are the so-called "Golden Tricolors." They require a rich light soil, such as fuchsias
would grow luxuriantly in, and should be planted out last
among the zonals. It is too much the custom to spoil these
plants by coddling them. The whole of the bedding stock
should be planted out, and the whole of the cuttings should
be struck in open borders. As, however, they are slow in
making roots, it is best to begin with these in June, and to
keep them slightly shaded, and regularly sprinkled, until they
have made roots. Some curious reader may ask, "How shall
I know when they have made roots?" Easily enough.
Instantly upon cuttings putting out roots they begin to grow,
and when new growth begins, the plants may be, comparatively speaking, neglected for awhile. As it is often a matter

of some importance to multiply these plants by every possible means, it is necessary that the reader be initiated into three great mysteries. Cuttings may be struck all the year round under glass. For this purpose, make up a bed in a greenhouse, the materials of the bed to be equal parts of sharp river-sand, and cocoanut-fibre refuse. In this mixture the smallest bits of stem, providing they have each a good healthy leaf, will soon make roots. From March to October no heat will be needed, but in the remaining four months the beds must be heated. Therefore it is well to make a bed for winter work in a proper propagating house, or on the top of a tank connected with the "flow" of the hot-water pipes. This brings us to mystery the second. It is quite a common thing for cuttings of tricolors to "damp off" in winter; in other words, to rot away instead of making roots. To prevent this, proceed as follows. Take two small pieces of stick, say small worn-out wood tallies for example, tie one of them across the cutting, about the sixteenth of an inch above its base, with a strip of bast or worsted. Tie the other lengthwise to the cutting, so that it projects two inches beyond the base, and overlies the crosspiece. Now, if the upright stick is thrust into the earth until the base of the cutting just touches the soil, the cutting will be held firmly in its position, and in due time will throw out roots, which may be covered with a sprinkling of the mixture the bed is made of. By this mode of procedure an enormous number of soft shoots may be struck during winter, and the losses by damping will be "next to nothing." The third mystery may be disposed of in a word. All the tricolors grow more rapidly when grafted on the common zonals than when on their own roots. Graft at any time from March to August, always keeping the plants extra warm, and somewhat shaded for a month afterwards. The best stocks are common seedlings. The following are the best varieties for bedding: *Victoria Regina, Louisa Smith, Sophia Cusack, Macbeth, Beautiful Star.*

Silver Zonals are generally known as "Silver Tricolors." They should be grown in the poorest ground, and, if possible, in raised beds. If grown in rich and rather damp soil, the leaves grow to a large size, are much wrinkled, and the dark zone is so fully produced as to spoil the effect of the variegation. In all other respects treat as advised for the Golden

Zonals. The best for bedding are *Imperatrice Eugenie, Queen of Hearts, Italia Unita.*

Bronze Zonals are, for the most part, vigorous growers, and a few of them are attractive as bedders. Many that are extremely fine in pots become either too green or too brown when planted out, and, therefore, it is important to select them with judgment. The best are *Downie's Princess of Wales, Imperatrice Eugenie, Countess of Kellie, Waltham Bride, Mulberry Zone, Egyptian Queen, Duke of Edinburgh, Mrs. Lewis Lloyd.*

Golden Selfs.—These are the most valuable of all the varieties for bedding where a distinct yellow or sulphur-green is required, as they present, in the mass, only one tone of colour, whereas the golden and silver zonals (tricolors), and the bronze zonals (bicolors), tend more or less to produce a mixed effect, wanting in unity and decisiveness. It must be admitted that a good bed of either of the classes just named is most beautiful when we stand near it and look down upon its rich mosaic of colours; but for a more distant view and for a distinct chromatic effect the golden selfs are unsurpassed, and are especially valuable for leaf embroidery, if the trusses are constantly pinched out before the flowers open. In this section the following are splendid bedding plants: *Meridian Sun, Golden Glory, Crimson Banner* (this has lovely magenta coloured flowers, and makes a remarkably rich bed if it can be allowed to flower), *Jason, Golden Fleece, Little Golden Christine.* The last is a miniature plant suitable for edgings.

Golden Edged.—These have a more distinct green disk than the golden selfs, in which the disk is so inconspicuous that we regard it as non-existent. The two classes might, indeed, be fused into one, because it is impossible to draw a sharp line between them; but the division is convenient, and is founded on degrees. The best in this class is a very fine old variety, which many cultivators condemn because they cannot grow it. But those who can manage it know it to be invaluable. Perhaps the mention of *Golden Chain* may carry many a reader back to pleasant remembrances of scenes and circumstances in days gone by, when the bedding system was in its infancy, and the hand that now holds the book was firmer in its grasp and readier for action in outdoor industry than now. But gushing is not allowable in a work of this sort, and so we quit the "pleasures of memory" to remark

that Golden Chain, in common with many other slow-growing
varieties, should be taken care of, to secure a good stock of
old plants, for this variety cannot be considered quite suitable
for bedding until the plants are three years old. It is but
proper to add, however, that vigorous growers like Meridian
Sun and Golden Banner are remarkably effective in their first
season, but, nevertheless, the enthusiast in parterre colouring
will never regret the exercise of patient care in the proper
treatment of this fine old favourite of the garden. Add to the
stock of golden-edged geraniums, *Crystal Palace Gem*, which
is almost a self, *Gold Circle, Creed's Seedling*, and *Yellow Gem*.

Silver Edged.—In this class we place all the white and
creamy-toned "variegated geraniums." They are rather
delicate in constitution, and old plants are to be valued, especially if judiciously cut down to keep them dwarf and
bushy. The best of the whitest are *Flower of Spring, Silver
Chain, Queen of Queens, Bijou, Snowdrop*, and *Avalanche*.
The last-named has white flowers, and, therefore, the flowers
need not be removed. The best of the creamy-edged are
Daybreak, Oriana, and *Flower of the Day*.

Green Ivy-leaved geraniums are useful for edgings and for
baskets. The best are *Bridal Wreath, Gem of the Season*, and
Willsi rosea.

Golden Edged Ivy-leaved geraniums make lovely edgings
where they can be employed with advantage. The common
Golden Ivy-leaved or *Aurea variegata*, which is its grand name,
is quite a gem in its way. The other varieties of this section
are all second-rate.

Silver Edged.—The best of these are *L'Elegante* and *Silver
Gem*.

Hybrid Geraniums, in poor soil, may be planted out ; but
where the soil is strong it is best to plunge them in pots, and
have a reserve of plants to take their places in case they fail
before the season is over. As for the reserve plants, the way
to insure having them in bloom when they are likely to be
wanted is to cut them back in May, shift them early in June,
at the end of June pinch out the tops and all the trusses,
and then let them push their trusses to be ready for service
in the parterre. The *White Unique, Crimson Unique*, and
Purple Unique are splendid plants. *Bridal Ring, Britannia*,
and *Ignescens superba* are, in their way, extremely useful.

PENTSTEMON.—For large isolated beds the garden varieties

are grand furniture for autumn display. A rich deep soil is requisite, with full exposure to the sun. The best way to manage the stock is to strike a sufficiency of cuttings in frames about the end of August or early in September, and keep them rather dry through the winter, to plant out in April. Treat the same as calceolarias in fact, but strike the cuttings earlier. The following are splendid bedding varieties: *Agnes Laing, Arthur McHardy, Miss Hay, Mrs. Sterry, Shirley Hibberd, Stanstead Surprise.*

PETUNIA.—This old favourite is now but sparely planted in the parterre, but it has certainly not been superseded, and in the hot summers of 1868 and 1870, a few of the varieties were remarkably showy, and held their own bravely to the very end of the season. For hot dry soils and in hot dry seasons the petunia is invaluable. In rich soils and in moist seasons it does not flower freely, and it grows too rank and green to be valued as a bedder. It is an easily-managed plant, provided the stock is wintered with care in a dry airy house, always safe from frost, and with no more water than just suffices to keep it green until spring returns. The usual plan of multiplying is by cuttings, and the best time to strike them is in the latter part of March and early in April. They may be struck as late as May, and will, with proper care, make good plants to begin flowering in July. The best amongst a thousand for bedding is *Spitfire*, a brilliant purple flower. *Shrubland Rose, Crimson Bedder, Purple Bedder*, and *Magnum Bonum*, afford a sufficient selection of single varieties for all ordinary purposes. The double petunias make fine pot plants, but are of quite secondary importance as bedders. The best of them for outdoor display are *Miss Earl*, rose, with white centre, and *Princess*, dark crimson. A few of the most useful bedding varieties reproduce themselves very faithfully from seed, if due care is exercised in saving it. The two varieties that can be depended upon most in this respect are *Countess of Ellesmere*, rose, with a light centre, and *Prince Albert*, deep purplish crimson. A bed of striped varieties or of mixed sorts, such as white, purple, and rose, has a very charming appearance, but of course they are not suitable for a bed that occupies a distinctive position in a geometric scheme. Petunias should be grown in rather poor soil, as they grow too luxuriantly and become coarse in soil enriched with manure. It is also essential to edge the beds with some strong-growing subject, such as *Centaurea ragusina*, or, what is better still, strong two-year-old plants of variegated pelargonium *Bijou* or other erect-growing variegated varieties. A solid edging, as here suggested, will keep the growth of the petunia in its place, and the beds will have a neat appearance, if the young growth is not allowed to ramble through the edging.

PERILLA.—Though the popularity of this plant has greatly declined within the past few years, it cannot be dispensed with, for its solemn bronzy-purple colour gives it a most distinctive character, of great value to the colourist. It has been well abused for its " funereal " aspect, and greatly misused by planters, who, in common with its detractors, were ignorant of its capabilities and proper place in the disposition of colours. The plants are always raised from seed in the first instance, but the tops may be taken off in June and July, and struck in about ten days in a frame, if a further supply is required for planting in the autumn. The middle of March is early enough to sow the seed, and a very mild heat suffices for its germination. As a rule, the seed is sown too early, and the plants, during the early stages, grown in too much heat. Stocky plants, three inches high at planting time, are decidedly preferable to gawky things eight to ten inches high with a few leaves at the top only. Of course, when required for a back row, it is necessary to have them rather tall, but they should be managed so as to insure being furnished with foliage to the surface of the soil. Perillas transplant so well that it is not necessary to put them in pots, and very satisfactory results may be obtained by pricking off from the seed-pan into a bed of soil made up in a cold frame. There are several varieties, but *P. Nankinensis* is the best.

PHLOX.—The large-flowering phloxes are not adapted for the parterre, but the varieties of *P. Drummondi* are invaluable for their continuous bloom and brilliant colours. It is usually supposed that a bed of phloxes must be mixed, but that is a mistake, for the named varieties come sufficiently true from seed, and a few of them deserve to be regarded as amongst the best bedding plants in cultivation. It is quite a waste of labour to plant them in a hungry soil, or to allow them to perfect seeds, for in either case they will present a shabby appearance long before the summer is past. The soil in which they are put when removed from the seed-pans should be moderately rich, to insure a healthy growth during the time they are in pots. The most distinct and valuable varieties are *Alba*, pure white; *Atrococcinea*, deep scarlet; *Heynholdi*, a magnificent scarlet-flowered variety; *Leopoldi*, pink and white; *Queen Victoria*, dark rosy purple. A bed of mixed sorts has a fine appearance, but seed of each of the above should be procured, and the plants arranged according to their colours when planted. The seedlings must be potted off immediately they are strong enough, because it ruins them to be allowed to remain crowded together in the seed-pots after they are an inch or so in height. Sow in a gentle heat in the latter part of March or early in April.

PORTULACCA.—For dry and hot positions these are most useful. They are very singular in appearance, and the flowers are remarkably showy. The soil in which they are planted should be light and sandy. The most effective for bedding purposes are *P. alba*, white; *P. caryophylloides*, white and rose; *P. coccinea*, scarlet; *P. splendens*, crimson; *P. Thorburni*, yellow. Sow in pans of sandy soil in April, and instead of putting the seed-pans in heat, place them on a sunny shelf in the greenhouse, and lay a slate over until the little plants appear. Pot them off early and keep them in a dry sunny position until they are put out in the flower garden. A few groups planted upon a rockery having a sunny aspect would produce a startling effect.

PYRETHRUM.—The great usefulness of the *Golden Feather* is too well known to require a single word of praise, and very few words will suffice respecting its management. As a rule, amateur gardeners sow much too early, and place the seed-pans in a strong heat. The end of March is quite early enough to sow, and the temperature of a cold frame will be quite sufficient. Seed may be sown in the beds about the beginning of April, but it is more advantageous to sow in a frame and transplant. The frame should, of course, be well ventilated after the seed has vegetated. Sow thinly in drills, and then the whole stock can be transferred direct from the seed-bed to the flower garden. As a matter of course it must not be allowed to flower in the parterre, but for the purpose of saving seed a few plants may be put out in some odd sunny corner or in the kitchen garden.

TROPÆOLUM.—It may be said with safety that the value of the tropæolum for flower garden decoration has been overrated, and also that, although a very large number of new varieties have been sent out within the last three or four years, very little real improvement has been effected. Two-thirds, at least, of the newest varieties are worthless. The varieties of *Tom Thumb* are of no real value in the parterre, as they seldom, if ever, bloom continuously throughout the season. They make a brilliant display for a short time and then go out of bloom, and remain an eyesore for the remaining part of the season, unless pulled up and consigned to the rubbish-heap. A few ·clumps placed in a mixed border, where they can be pulled up as soon as they begin to present a shabby appearance, are very well, but they should not be planted extensively. The only recommendation they have consists in the fact that they reproduce themselves freely from seed, and the only trouble occasioned by raising a stock consists in sowing the seed where the plants are to remain and flower. The well-known *T. Lobbianum* and its varieties should be planted in poor soil, in the most sunny position, to insure an abundant bloom. In the event of the soil proving too strong, the plants acquire a coarse leafy character, which it is impossible to correct in a satisfactory manner, though frequent removal of the leaves, where they are crowded, will reduce the luxuriant habit of the plants and promote a more free production of flowers. The soil, however, should be deeply dug to encourage the plants to send their roots abroad, and enable them to hold their own against a drought. It is also important to put out strong plants that are well hardened off, for when they are not more than half hardened previous to planting, they generally receive so very much injury from the sun and wind that the summer season is far advanced before they become well established. The stock of all the bedders should be renewed by cuttings, for seedling plants cannot be depended upon, as all vary more or less, either in the colour of the flowers or the character of the growth. The best of the scarlet-flowered varieties are *Beauty of Malvern* and *Star of Fire*. Both are neat and compact in growth, and bloom most profusely throughout the season, and, unlike the Tom Thumb varieties, they do not produce much seed. *Advancer* has also scarlet flowers and is very desirable, and in some soils may, perhaps, equal in effectiveness both the foregoing. Of the varieties producing flowers other than of a scarlet hue, *Luteum Improved*, rich orange-yellow spotted with crimson, and *The Moor*, deep crimson-maroon, are the best.

Though the *Tom Thumb* section, which are improved varieties of the "dwarf nasturtium" of days gone by—the
Tropæolum minus of botanists—have been referred to above as
"of no real value in the flower garden," we must bestow a few
words upon them, because they still retain a shadow of the
favour with which they were regarded in the early days of the
bedding system. With all their faults they are extravagantly
showy while they last, and may be employed to advantage in
beds that are to be managed on the "chameleon" principle,
the object of which is to present in one and the same spot a
succession of masses of colour throughout the season. It is
a good plan to put them rather far apart in the beds, and
plant some of the tall-growing asters between them. The
asters will not produce such fine flowers as when planted in a
bed by themselves, but they will flower freely and take the
place of the Nasturtiums as soon as they begin to present a
shabby appearance. To prolong the flowering season as long
as possible, the seeds, before attaining half their usual size,
should be picked off, because the plants will certainly not
flower freely if they have to perfect a crop of seeds. The
seeds will pay for the cost of gathering, as they make a most
acceptable pickle when gathered green. The best varieties are
Scarlet King of Tom Thumbs and *Golden King of Tom Thumbs.*
The seed should not be sown until the end of March, or the
plants will be too forward. The simplest and best way of
dealing with them is to sow four or five seeds in five-inch
pots, and then thin the plants down to two or three to each.

VERBENA.—There is not in all the catalogue of bedding
plants one that more perfectly answers to the requirements
of the garden colourist than this. Its trailing habit, forming
a close carpet of vegetation, its well-sustained umbels of
brilliantly-coloured flowers glittering above the suitable
groundwork of dark green leaves; and the continuousness
of its intensity of colour, are qualities that will insure it a
place in the select list of first class parterre plants. And yet
the verbena has been steadily declining in popularity during
many years past, in consequence of frequent failures, and the
consequent disfigurement of the gardens where it has proved
unequal to the demands and expectations of the cultivator.
It must be confessed that in exceptionally hot, dry seasons
like those of 1868 and 1870, verbenas unhappily situated,
shrink away to dust ere the season is half gone. It must be

confessed, also, that a large proportion of the newest varieties have been recommended for bedding, and have been tried and found wanting, to the injury of the fair fame the verbena should enjoy, and the actual discouragement of those who are labouring to improve it. Having made these admissions, it remains to be said that, as a rule, failures with the verbena result from bad cultivation, and especially of the careless system of planting bedders in badly-prepared soil, without, in any case, any special preparation for any of them.

It is only in a good deep holding loam that the verbena will grow in a satisfactory manner; but a light soil will suit the plant, provided a liberal dressing of manure is dug in during winter, and a fair average season follows, with alternations of showers and sunshine, for with the best preparation, a failure may be expected in a peculiarly hot and dry season on light sandy soils. As we do not often experience the delights and trials of a tropical summer, those who appreciate this plant may reasonably reckon on success in cultivating it, even though they may have a lighter soil to deal with than the plant would prefer, provided they adopt a liberal system of cultivation. In the case of a hot soil, a mulch—that is, a surfacing of half-rotten manure put on at the time of planting—will do wonders, and as to its appearance, the plants will so soon spread over and hide it, that it is practically of no consequence. In a droughty summer, a few heavy soakings with soft water will also act beneficially; but it is best to avoid watering if there is a prospect of rain before the plants begin to suffer, and, in any case, frequent surface dribblings do more harm than good. It is not a matter of great importance to plant verbenas in the full sun, but a heavily shaded position will not suit them. A free current of air, and a few hours of sunshine per diem they must have, but they cannot so well endure continuous roasting as geraniums and petunias, which really rejoice in sunshine. A very common cause of failure is the practice of putting out plants that have been starving in small pots several months previous to the planting seasons. It is a grievous mistake to propagate the stock for bedding early in the season, although it is generally supposed that early propagation is necessary to secure strong healthy plants by planting-time. When struck early, and necessarily kept starving in pots for several months, the constitution becomes impaired so much that they are unable to resist, with any degree of success, the attacks of red spider, thrips, and mildew—three most formidable enemies they have to contend with. The month of April is quite early enough for striking verbenas intended for bedding purposes. The tops of the healthy shoots should be taken off in the early part of the month, struck in a brisk bottom-heat, and potted into store pots, and carefully hardened off; these planted out as early in May as the weather will permit, will grow away freely, and the beds in which they are planted will soon become a blaze of colour. The compost in which they are potted should be rich and nourishing, and for that reason nothing suits them better than a mixture of good turfy loam and decayed hotbed manure, mixed together, at the rate of two parts of the former to one of the latter, and a sprinkling of sand added to keep the compost open.

In a collection of over a hundred kinds grown in our experimental garden in the burning summer of 1870, the following were the best:—*Annie*, a free-flowering variety, prettily striped. *Ariosto Improved*, rich puce or plum-colour. *Blue King*, light blue, distinct and pleasing. *Crimson King*, fiery orange-scarlet with small lemon eye ; the best scarlet-flowered verbena we have for bedding purposes. *Firefly*, fine brilliant scarlet, very showy. *Grand Boule de Neige*, pure white, very large. *Iona*, rich crimson, very dwarf. *Isa Eckford*, rich puce. *King Charming*, clear salmon rose, distinct and showy. *Lady Folkestone*, deep rosy purple. *Madame Lefebvre*, bright reddish crimson. *Mrs. Eckford*, white with rose centre. *Mrs. Reynolds Hole*, white with crimson centre. *Mrs. Pennington*, rich reddish rose. *Murillo*, shaded peach ; contrasts well with the crimson, puce, and other dark colours. *Otago*, brilliant rosy crimson, worthy of a place in the most select collections. *Parsee*, bright purple flushed with mauve. *Polly Perkins*, bright rosy red. *Purple King*, an old but most valuable variety. *Reine des Roses*, deep rosy pink. *Storm King*, rich rosy crimson. *Victory*, brilliant scarlet, strong in growth.

VIOLA.—A few pretty violas have been turned to account as bedding plants, and have, in this capacity, acquired far more fame than they deserve. For flowering in spring and early summer they are invaluable, but for summer and autumn comparatively useless. The named varieties should be grown from cuttings in the same way as recommended for pansies.

When raised from seed, sown in March or April, a better autumn bloom may be expected, than from plants raised from cuttings. The varieties of *V. cornuta* have blue flowers. *Perfection* is the best of them. *V. lutea* is a good yellow-flowered species, well adapted for a display in spring.

A SELECTION OF PLANTS SUITABLE FOR EDGING FLOWER BEDS.

SILVERY-LEAVED.—*Achyrocline Saundersoni*, 6 to 8 in., neat and upright. *Arabis albida variegata*, forms a close tuft of creamy-leaved herbage; hardy. *Achillea umbellata*, 4 in., bushy and neat. *Antennaria tomentosa*, 1 in., spreading; hardy. *Centaurea ragusina compacta*, 6 in., bushy; nearly hardy. *Cerastium tomentosum*, 3 in., spreading; hardy; requires clipping into shape two or three times during the season. *Cineraria acanthifolia*, 6 to 12 in., upright, bushy, large beds; *C. asplenifolia*, 9 to 12 in., bushy, large beds; *C. maritima*, 9 to 15 in., bushy, large beds: all from seed or cuttings. *Dactylis glomerata elegantissima*, 9 in., bushy; hardy. *Echeveria glauco-metallica*, 4 in., neat, fine for sloping sides; *E. secunda glauca*, 2 in., sloping sides of large or small beds, very neat. *Gnaphalium tomentosum*, 6 to 18 in., branching; requires clipping into shape; *G. lanatum*, 9 to 15 in., straggling in growth, and requires to be clipped frequently. *Euonymus radicans variegatus*, 6 to 12 in., can be clipped into shape; medium-sized or large beds. *Polemonium cæruleum variegatum*, 6 in., very compact; light soil. *Senecio argenteus*, 3 to 5 in., neat bushy rosettes; hardy and very valuable. *Stachys lanata*, 6 in., coarse in growth, and increases fast; should be taken up and divided every year. *Veronica incana*, neat and compact; large or small beds; requires dividing; hardy and very valuable.

GOLDEN-LEAVED.—*Arabis lucida*, 4 in., neat; sandy soil; increased readily by division. *Aucuba-leaved Daisy*, 3 in.; beautiful during winter and spring. *Euonymus flavescens*, 6 to 18 in., rich chrome yellow; hardy. *Fuchsia Golden Fleece*, 6 to 8 in., bushy and compact. *Golden Feather*, 3 to 6 in., bushy and compact; should be raised from seed. *Lonicera aurea reticulata*, 9 to 15 in., fine for large beds; must be pegged down. *Mesembryanthemum cordifolium variegatum*, 3 in., spreading; beds and borders in hot situations. *Thymus citriodorus aureus*, 3 in., spreading, but compact; hardy.

CHAPTER VI.

HARDY BORDER FLOWERS.

THE hardy herbaceous border is the best feature of the flower garden, though commonly regarded as the worst. When well made, well stocked, and well managed, it presents us with flowers in abundance during ten months out of twelve, and in the remaining two blank months offers some actual entertainment, and many agreeable hints of pleasures to come, to make an ample reward for the comparatively small amount of labour its proper keeping will necessitate. Given a few trees and shrubs, a plot of grass, and comfortable walks, the three first essentials of a garden, and a collection of hardy herbaceous plants is the fourth essential feature, and may be the last; for the bedding system may very well be dispensed with in a homely place, provided the hardy flowers are admitted, and cared for, according to their merits. It may be that many a reader of this will be disposed to question whether geraniums should be swept away to make room for lilies, and verbenas denied a place because of the superior claims of phloxes, but such a question we do not propose—our business is to point out that the bedding system is an embellishment added to the garden: the herbaceous border is a necessary fundamental feature. Therefore we ask for the establishment of a collection of herbaceous plants before preparations are made for a display of bedding, and our advice to those who love their gardens and walk much in them, and find amusement in watching the growth of plants, and in contrasting their various characters and attractions, is, that they should seek to develop the herbaceous department, and so become acquainted with its full capabilities. In this pursuit enthuriasm may be manifested without incurring the reproach of season, for it is a truly intellectual pastime, and demands the practice of patience, and the exercise of thought in no small

measure from those who would know more of it than appears
upon the surface. Let us for a moment consider the claims
of the herbaceous border to better regard than is usually
bestowed upon it.

It is an important characteristic of the herbaceous border
that its proper tenants are hardy plants that need no aid of
glass or fuel for their preservation during the winter. Those
who can be content with hardy plants alone may find it an
agreeable and easy task to devote their glass-houses to the
production of grapes, mushrooms, forced kidney beans, and
other equally valuable delicacies, and supplement the hardy
garden with a collection of Alpine flowers, a large number of
which can be better grown and more thoroughly enjoyed in
an airy and unheated greenhouse than when planted on the
rockery in the open air. The delights of spring may thus be
antedated by the aid of glass, and suitable early-flowering
Alpine plants and the open borders will present an abundance
of flowers, from the time when the treacherous frosts have
spent their spite upon vegetation until the chill of winter
returns again. In the cultivation of bedding plants we may
fairly reckon on a brilliant display for three months, and it
may extend to four—say, from the 1st of June to the 30th of
September, but the herbaceous border will be gay from the
end of April to the middle of October, a period of six months,
and will offer us a few flowers in February, and a few in November and December, and in a mild winter will not be utterly
flowerless even in January. It would be an exaggeration
to say that the herbaceous border is capable of a display of
flowers all the year round, but it is very nearly capable of a
consummation so devoutly to be wished. To the advantages
of hardiness and continuity of bloom must be added a third
and grand qualification, of a distinguishing kind—that of
variety. It is scarcely an exaggeration to say that the
varieties of form, colour, and general character, amongst
hardy herbaceous plants is without limit; but, as variety may
be obtained amongst ugly plants, we are bound to add that
the proper occupants of the garden we are considering are all
beautiful, and a considerable proportion are well known
favourites. Nevertheless, it must be admitted that with all
their good claims to loving regard, the hardy herbaceous
plants obtain but scant attention, and tens of thousands of
persons who know that verbenas are somewhat showy when

in flower, and would like to grow thousands of them, are prepared any day to ignore the whole tribe of herbaceous plants as weedy things that have had their day, and, with the exception of a lily or two, and, perhaps, a hollyhock, deserving of a place only in the unsavoury hole where grass-mowings and the sweepings of the poultry-house are deposited with a view to a "mixen." It ought to be needless to attempt this vindication, but we feel bound in duty to the reader to urge that every rational development of the hardy garden will prove advantageous to the lover of flowers, as tending both to lessen the expense and labour which the keeping of the garden necessitates, and considerably augment the pleasures that it is capable of affording as the seasons change and the year goes round.

As hardy herbaceous plants of some kind or other will grow in any soil and any aspect, not one single square foot of ground in any garden need be utterly barren. A tuft of Solomon's seal in a dark spot where the soil is quite unfit for better plants, may be better than nothing. Sunny, shady, hot, cold, dry, moist, or even wet positions, have their several capabilities for hardy plants, and we have but to make our selections prudently to insure a plentiful clothing of herbage and flowers for every scene. But a herbaceous border designed for a good collection of plants should consist of good deep loamy soil; the greater part of it should be fully exposed to the sunshine and the breezes, but it is well to have some extent of ground partially or considerably shaded, to provide the greatest possible variety of conditions for the greatest possible variety of the forms of vegetation. In preparing a border, in the first instance the ground should be well dug two spits deep and at the same time liberally manured. In the case of an old border requiring a repair, it may be well to lift all the plants and "lay them in" safely while the border is trenched and manured; or it may suffice to leave the good plants undisturbed and provide sites for additional planting by opening holes and digging in plenty of manure. In any case we would earnestly advise that herbaceous plants should be thoroughly well cultivated, even if, to do full justice to them, the bedding display has to be contracted or abolished. The majority of the best herbaceous plants—the hollyhocks, phloxes, lilies, tritomas, delphiniums, pinks, chrysanthemums, primulas, pyrethrums,

potentillas, anemones, ranunculuses, irises, œnotheras, fox-
gloves, campanulas—require a deep, rich, well-drained loam,
but will grow well in clay that has been generously prepared, and need not be despaired of altogether where the
soil is shallow and sandy, provided there are appliances available in the shape of manure, mulchings, and waterings, to
sustain them through the hottest days of summer. It must not
be forgotten, too, that if the herbaceous border is formed on a
somewhat good soil—say a soil that will grow a cabbage—and
in a position open to the sun and the health-giving breezes, it
may be enriched by the addition of roses, stocks, asters,
zinnias, balsams, dahlias, and many more good things, that
"need only to be seen to be appreciated."

In the management of the herbaceous border details are
everything, and principles next to nothing. The best time to
plant is in August and September, but planting may be safely
done in March and April, and with but little risk on any day
throughout the year, provided the plants at the time of planting are in a proper state for planting. For example, a hollyhock may have a spike of magnificent flowers six feet high in
the first week of September, and no sane gardener would then
propose to transplant it; but the white lily, only a yard or so
distant from it, may be just then in a dormant state, and, if
to be transplanted at all, in a condition most desirable for
the process. A great tuft of Arabis might be lifted any day
from October to February, if lifted quickly and replanted
with care, and in the ensuing month of April would bloom as
well as if left undisturbed; but any sensible person who had
struck a lot of arabis cuttings in pots in autumn would take
care not to plant them until May, because little weak scraps
of plants would probably perish if planted in the dark, short,
cold days of the year! Leaving a fair margin for exceptions,
it may be said with truth that herbaceous plants may be planted
at any time, but we must return to the primary presumption,
and repeat that the best time is in August or September, but
if the chill November days occur before the work can be done,
it is better to wait until spring, and then if possible choose a
time when the wind is going round to the west and the
barometer is falling. Haply, when your work is completed,
soft showers will fall to help your plants make new roots
quickly, to hold their own through the summer heat.

It cannot be wrong to repeat that the amateur need not

7

be troubled about principles, but must consider the management of the herbaceous garden a matter of detail. As to
watering, never give a drop if you can help it; but if it must
be given, give plenty. Plants that have a deep well-manured
bed to root in will rarely need water; but in some hot dry
places watering is a necessary part of the routine management
of a garden, and the herbaceous plants will be as thankful
as any for whatever help the water-pot can give them. Some
plants require stakes and some do not. Those that need
support against wind should have it in time, for the storm
may come and blow down half your garden wealth on the
very day you have begun to talk of staking the dahlias and
hollyhocks "to-morrow." We are no advocates of scanty
planting; we rather prefer a crowded garden, but must contend always for a sufficiency of the comforts of life for all
kinds of plants. The subjects we have before us require a
deep nourishing soil, and plenty of light and air, which over-
crowding will simply prevent them having; but a meagrely-
planted border has as miserable an appearance as a great dinner
table with only half a red herring on it. Always plant enough
to make a good effect at once, and in a year or two afterwards
thin out and transplant, or give away, or sell; don't waste
years in the expectation that you may obtain from half-a-
dozen plants enough stock to cover an acre, because it is not
well to make a nursery of a garden, and a good stock of all
the best things that can be obtained will afford far more
gratification than any quantity of some half-dozen sorts that
you may any day buy at about a fifth, or, perhaps, a tenth, of
what you must expend to produce them. Herbaceous plants
are, for the most part, easily multiplied, and, generally speaking, may be increased by the very simple process of division ;
but it is better to plant a small plot of ground in such a way as
to insure a good effect at once than to lay out a great extent
of border space with the intention of filling it "some day"
with home-grown stock. To enjoy herbaceous plants they
should be left undisturbed for years, to form great masses or
"stools," as they are called, for it is only when thoroughly
established that many of the best of them present their flowers
profusely and show all their characters in full perfection. It
is a strange thing that people who are always ready to expend money in the most liberal manner on bedding plants
become ludicrously niggardly the instant they become con-

vinced of a glimmering of faith in herbaceous plants. An
instance of this has amused us lately. When inspecting a
stock of hepaticas in flower in Ware's great nursery at Tot-
tenham, we met a customer who was enraptured with them.
Having, in company with some half-dozen persons, enjoyed
the brilliant display of colour produced by some three or four
thousand plants in a mass, this admirer ordered one plant,
which, being drawn out at once, was found to consist of a
tuft as large as a duck's egg, with two flowers expanded, and
three or four leaves on the way. The attendant naïvely suggested that people should buy these things in the same way
that they buy bedding plants—by the dozen, the score, the
hundred.

The best way is not everybody's way. The furnishing of
an extensive border by the purchase of sufficient of the very
best herbaceous plants, will prove a more expensive business
than every reader of this book may be prepared for. It
follows that something should be said on the raising of plants
by cheap and simple methods of procedure. Many good
plants produce seed abundantly, and the careful cultivator
may by this means increase the stock to any extent that may
be desired. The best seed is that saved at home, and the
best way to deal with it is to sow it, as soon as it is ripe, in
large shallow pans and boxes, and keep these in cool frames
until the plants appear. Some kinds of seeds remain a whole
year in the soil before they germinate, and therefore it is only
the patient who are well rewarded. As amateurs are apt to
lose seeds that they would fain save, we shall present our
readers with a rule of action that we have followed many
years in saving seeds of all kinds that are likely to scatter as
they ripen. Provide a lot of common bell-glasses, of various
sizes, and place them mouth upwards on a bench in a sunny
greenhouse. When a cluster of seeds is full grown and just
beginning to ripen, cut it and throw it into one of the bell-
glasses, with a label inscribed with its name. The ripening
process will soon be completed, and the seed will shell itself
out from the pods, and be found ready cleaned and fit for
storing away with the least imaginable amount of trouble.

We have saved all kinds of seeds in this way, and may say
with truth that the scheme has been worth hundreds of pounds
to us. The ingenious practitioner will soon discover how to
modify the plan advantageously. Thus, flower-pots, with the

holes stopped with corks or sheets of paper, may be used in place of bell-glasses; but the best way will pay the best, especially in the case of amateurs who grow "good things," and prize the seeds of choice subjects like gold-dust. We shall have to treat on seed-sowing and the management of seedling plants at length in a subsequent chapter, and therefore, to avoid waste of space by repetition, shall say no more upon the subject here. As to the value of seed-saving and seed-sowing, however, we are bound to repeat that in the case of herbaceous plants, the matter is not of the highest importance. How absurd, for example, it would be for any one to save and sow seed of the common white arabis, when, by the simple process of division in autumn, the plants can be multiplied *ad infinitum!* What a waste of time to wait and watch for seeds of the white lily, which only needs to be taken up and parted in August or September to fill the whole garden, no matter how large, in the course of a few years. It is worthy of remark, too, that, as a rule, the plants which produce abundance of seed are those that we prize the least; the free-seeding sorts being of secondary value as regards interest and beauty.

The multiplication of herbaceous plants by cuttings and divisions, when either of these methods can be practised, is far preferable to raising them from seed. The cuttings should consist of new shoots of the season, nearly full grown and just about to harden. Old and wiry shoots are of no use; very soft, sappy shoots are no use. Large cuttings, whether from old or young shoots, are no use. The mild heat of a half-spent hotbed is to be preferred to the strong heat in which bedding plants are struck in spring; but hardy herbaceous plants may be propagated in a strong heat, or a mild heat, or without heat, and the last mode is the best, generally speaking. In the case of a scarce and valuable plant, we must sometimes adopt extreme measures to save its life or to increase it rapidly; but the best plants will be obtained from the well-managed cold-frame, and not from the hothouse.

In multiplying by division, a time should be chosen when the plant is in what we may call a dividable state; but, in truth, it may be done at any season if the operator is somewhat experienced, and can coax an insulted plant into a kindly temper by good frame or greenhouse management. When we meet with a scarce plant that we wish to possess, we

secure, if we can, a cutting or a rooted shoot, or " a bit of it,"
somehow, and feel bound to make that " bit " a plant by some
means. Experience has taught us, in respect of scarce and
valuable plants, the best time to secure seeds, roots, cuttings,
offsets, etc., etc., is, *when you can get them,* and we know
nothing of seasons whatever. But in case this defining should
perplex an amiable reader, we shall wind up this paragraph
by saying that dividable subjects, such as violets, pansies,
daisies, arabis, and primulas, should be taken up in August or
September, and be pulled to pieces and replanted immediately.
If the weather is showery, they will prosper without any particular attention; but if the weather is hot and dry, they must
be watered and shaded until the cool, damp season returns.
It is a good plan to have a plot of reserve ground in which to
plant out the young stock, and allow it to make one whole
season's growth before transferring it to the borders.

Many disappointments occur through mixing tender and
hardy plants together in borders, and leaving them all to settle
accounts with the weather. They are very straightforward in
their mode of settlement. The hardy plants live and the
tender plants die, and those who have to pay for the losses
make long faces when summer returns and the favourites of
the past season are seen no more. In very severe winters,
and especially in gardens in valleys where the soil is heavy
and damp, many plants, reputed hardy, are sure to perish.
Losses are always objectionable; but a certain number must
be borne with in every pursuit, and the herbaceous border
forms no exception to the general rule. But the fact suggests
that a systematic use of frames and other like protective
agencies, and a reserve of plants of kinds that are least likely
to suffer by severe weather, are precautions the wise will
adopt without any great pressure of persuasions.

To speak of our own case for a moment, we cannot keep
hollyhocks in the borders during winter, and therefore take
cuttings in time, and secure a good stock of young plants in
pots in autumn, to keep through the winter in frames for
planting out in the month of April ensuing. The amateur
must study these matters as essentials to the realization
of the true joy of a garden. Borders that are kept scru-
pulously clean all the winter will be the most severely
thinned of plants in the event of extra severe weather.
There is no protective material so potent to resist frost as the

dead and dry leaves of trees, as the wind disposes them, for
they always gather about the crowns of herbaceous plants,
to help them through the winter.

After winter comes the spring, and then the gardener
will carefully dig the border, and chop up the roots of pæonies,
and stamp down with his foot the pushing crowns of anemones, and by a most unavoidable accident chop up a few of
the phloxes. We never suffer the herbaceous border to be dug at
all, except to prepare it for planting in the first instance, or for
needful repairs afterwards. Periodical digging, "as a matter
of course," such as the jobbing gardeners designate "turning
in," has for its sole object the destruction of plants; but
that object is disguised by describing the operation as
"making things tidy." When you are tired of herbaceous
plants, let the jobbing gardener keep the border tidy, and
you will soon soon be rid of the obnoxious lilies, phloxes,
ranunculuses, anemones, hollyhocks, pæonies, and pansies,
without the painful labour of pulling them up and burning
them.

CHAPTER VII.

It must be understood that in this selection we can have nothing to do with curiosities, or with plants that are simply "interesting" for some odd reason to somebody. We must have beauty, or at the very least a somewhat showy character, in every plant selected. There are, perhaps, fifty species and varieties of lilies known to gardeners, but about half a dozen are enough for any amateur who has not committed himself to the idea of living on lilies and living for lilies at any sacrifice to the end of his days. Better is it, we believe, to have some fine clumps of such comparatively common plants as the white lily, the orange lily, the golden-striped lily, and the ivory-flowered lily (e.g., *L. candidum, L. bulbiferum, L. auratum,* and *L. longiflorum*), than plant a lot of "curious" lilies that may cost a guinea a bulb to begin with, and be scarcely worth a farthing a bulb for their beauty when in flower, though some of the curiosities may require two or three years' growing before they deign to reward their patient owner with a hint of what they would be if they could. We earnestly advise the lovers of hardy plants to grow good things, and leave the bad things to the botanists. The herbaceous border must not be a refuge for weeds, labelled with hard names long enough to reach from here to the moon, but a comfortable home for beautiful flowers that need so little care that it may be said of them that the delight of owning them is not necessarily accompanied with the care of keeping them. It is not intended to name all the good things in the list that follows, but it is intended to include good things only, and we offer it as comprising a selection of the most beautiful herbaceous plants known to cultivation, comprising chiefly such as readily adapt themselves to diverse conditions of soil and climate.

ACHILLEA (Milfoil).—A quite unimportant group of plants. They will grow in any soil, and may be multiplied by division. *A. ægyptiaca* is a pretty white foliage plant, occasionally employed for edgings in the parterre. *A. filipendula* is a fine plant for the shrubbery, graceful in foliage, and with showy yellow flowers. *A. millefolium* is the common milfoil, a most valuable plant for lawns on dry hot soils, and for turfing banks. The variety with rose-coloured flowers is an extremely pretty shrubbery and cottage garden plant. The double flowering variety of *A. ptarmica* is a gem for the border, and a good plant to grow in pots for the conservatory, and moreover it forces well.

ACONITUM (Monkshood).—A showy family of rustic plants, of a most poisonous nature, which in any case should not be planted without consideration of the possibility of their proving dangerous. They are well adapted for large borders and the skirts of shrubberies, where their stately forms and handsome flowers show to great advantage. A deep, rich soil suits them well, and they will bear partial shade. They are propagated from seeds sown in spring, and division of their fleshy roots in autumn. The best are *A. napellus*, with blue and white flowers ; *A. Japonicum*, violet blue ; and *A. tauricum*, dark blue. The gigantic *A. lycoctonum* makes a striking object in woodland scenery, but cannot be considered a border plant.

ADONIS.—The best of this family is *A. vernalis*, an old favourite, with finely-cut leaves, and large yellow flowers, which appear in March and April. *A. apennina* is the same in character, but comes into flower immediately after vernalis. *A. pyrenaica* flowers in June. These plants require a deep moist rich soil. They may be increased by seeds sown in March, or by division of the root at the same season.

AGAPANTHUS (African Lily).—The well-known *A. umbellatus* is quite hardy, and is a first-rate border plant. It requires a deep rich moist loam, and will thrive equally well in sun and shade. Multiply by division of the roots in April or in August. The white-flowered variety is as hardy as the blue, but the variegated-leaved variety is scarcely hardy enough for the border.

AGROSTEMMA (Rose Campion).—The varieties of *A. coronaria*, of which there are at least three, are extremely showy, and have the good quality of flowering freely all the summer long. They will grow in any moderately good soil, and

prefer a damp or boggy situation, but must have a full exposure to the sun. Multiply by division and by cuttings.

ALLIUM (Onion).—Several species of garlic and onion are
worthy of a place in the best border, for they are most elegant
when in flower. Any soil will suit them, and they bear partial
shade without injury. They generally increase rapidly without attention by scattering their seeds when ripe; and therefore, if young plants are required, leave the soil undisturbed
around the old ones. The best are *A. album*, *A. moly*, *A.
roseum*, and *A. ciliatum*.

ALSTRŒMERIA (Chilian Lily).—These brilliant plants are,
with only one or two exceptions, perfectly hardy, and require
only the simplest cultivation. They are admirably adapted
for filling odd places and out-of-the-way nooks, where they
can be left alone undisturbed for years; that being one of the
conditions of success with them. The soil should be deep,
rich, and light, and it matters not whether the staple is peat
or loam, but it must be well drained, for if in the slightest
degree boggy, the winter will destroy the plants. Plant them
deep; give them plenty of water during summer, and in
winter cover with a thin sprinkling of tree leaves. They are
increased by divisions of the fleshy roots in autumn. The
best are *A. aurantiaca*, height 2 feet, flowers orange and yellow;
A. Errembaulti, 2 feet, flowers white, with crimson or yellow
spots; *A. psittacina*, 3 feet, flowers crimson and green. There
are many beautiful varieties in cultivation, in addition to the
three here recommended.

ALYSSUM (Madwort).—The well-known "Yellow Alyssum"
A. saxatile, makes such a brilliant show in the month of May,
that it is almost impossible to have too much of it. This
showy plant will grow in any soil, but requires an open
sunny situation, and is certainly somewhat unsafe if the soil
is more than ordinarily damp in winter. We have had to
grow thousands of it, and have always found cuttings of the
young shoots better than seeds, but it is easily multiplied by
either plan. The variegated-leaved variety of *A. saxatile* is
an extremely pretty plant for the rockery or for pot culture.
A. argenteum is a fine showy plant for the rockery, but of far
less value than the common alyssum for the border. The
"variegated alyssum" of the bedding system is *A. orientale
variegatum*, a decidedly tender plant of comparatively trifling
value.

ANEMONE (Windflower).—This is one of the most useful families, both for spring and autumn flowers. The species are all lovers of a deep, rich, moist soil, such as buttercups naturally take to. They bear shade well, and may be multiplied by divisions and seeds. *A. alpina* is a tall plant, flowering in April; the flowers large, creamy white inside, purple outside. *A. apennina* is a tiny plant, producing lovely blue flowers in March. It cannot be grown where snails and slugs abound; for they never cease to browse upon it while there is a leaf left. *A. nemorosa* is another sweet little gem, with pearly-white flowers; the double variety to be preferred. *A. rivularis* is a fine border plant; requires a damp soil, growing two feet high; flowers white. *A. sylvestris* grows a foot high, and produces charming white flowers in April. It is invaluable for the border. *A. Japonica* is another first-rate border plant, flowering from August to November. The common form has pink flowers, but there is a fine variety, with pure white flowers, named Honorine Jobert, which may be regarded as one of the most beautiful and useful border plants of its season. The Florist's Anemone, descended from *A. coronaria* and *A. hortensis*, both of which are fine border plants, is better known than the species above enumerated. The cultivation of named double anemones of the florists' section has of late years greatly declined, probably because considerable trouble must be bestowed upon them to secure fine flowers. They require an open position, and well-prepared, deep, rich, loamy soil. The roots are planted, in rows a foot apart and two inches deep, in October or November. If the soil in which they are grown is damp, it is advisable to defer planting until February; but they never flower so finely if planted in spring as if planted in autumn. The roots are taken up in May or June, and carefully cleaned and stored away in bags or boxes. Those who desire the showiest of anemones without the trouble of growing the double ones properly should plant in the border plenty of *A. coronaria* and *A. hortensis*. Of the latter, the varieties named *stellata, fulgens,* and *purpurea* are brilliant in colour, and make a fine display in spring. They should all be increased by division, unless the cultivator has some special object in growing them from seed.

ANTIRRHINUM (The Snapdragon).—*A. majus* is well-known for its gay flowers and its love of ruins. We see it flaunting its red and white banners on the top of the tower and on the garden wall, and are advised by the fact that it can live on little and spread itself abroad without the aid of man. To grow the plant from seed is indeed most easy, and all that need be said about it is that the seed may be sown at any time from March to September, and the best way to treat it is to sow in shallow pans in a cold frame, and plant out the seedlings in a bed of light earth in a frame, thus to remain during their first winter. When planted out in the following spring, a sunny, well-drained spot should be chosen, and although the plant needs but a mere film of soil to sustain it, a rich sandy loam will produce finer flowers, and more of them, than the handful of lime-rubbish on the top of the wall, where the vagrant snapdragon finds a lodging for itself. When grown from seed the flowers are various, and while some are pretty sure to be good, it is equally certain that many will be bad. Hence the named varieties of the florists' section are to be preferred for their distinct characters and splendid flowers. These are to be propagated by cuttings, which should be treated precisely as advised for the propagation of calceolarias at page 69.

BEST THIRTY ANTIRRHINUMS.

Acteon, Admiral, Artist, Bolivar, Bridesmaid, Bravo, Bella, Charming, Climax, Crown Jewel, Dr. M'Craken, Fire King, Flora, George Gordon, Gladiateur, Harlequin, Marquis, Mrs. M'Donald, Ne Plus Ultra, Orange Boven, Prince Alfred, Striata perfecta, Pretty Polly, Queen of Beauties, Queen of Crimsons, The Prince, The Bride, Undine, Wrestler, War Eagle, Yellow Gem.

AQUILEGIA (the Columbine) will grow in any good soil, especially if moist and rich, and will thrive almost equally well in sun or shade. All the species and varieties are worth growing, as they are neat and pleasing, and a few of them extremely showy. They are increased by division in autumn or spring and by seeds sown in March or April. Most of them sow their seed on the border, and soon form colonies in the same way as the antirrhinum. The most useful of all is the Common Columbine, *A. vulgaris*, of which there are many splendīd varieties, single and double. *A. alpina* is extremely pretty; the flowers are purplish blue. *A. canadensis* is a tall plant, with bright red and orange-coloured flowers. *A. cœru-*

lea is exquisitely beautiful, and one of the choicest herbaceous
plants known ; the flowers are of a delicate pale blue colour.
A. glandulosa is a showy species, with blue and white flowers.
A. Skinneri is a good one, with red and orange-coloured
flowers. They are all summer-flowering plants, making their
first display in May, and continuing to bloom until June or July.

ARABIS.—The Rock Cress presents us with one of the best
of all our spring flowers. *A. albida*, also known as *A. cau-
casica* and *A. crispata*. This plant forms a low-spreading tuft
of glaucous leafage, which in the month of April is completely
smothered with snow-white flowers. It will grow in any soil
and situation, but does not flower freely unless enjoying a
somewhat pure air and an open sunny situation. On a bank
or rockery consisting of sandy earth it acquires a glorious
luxuriance of growth, and should be allowed to spread if
space can be afforded it ; for though its season of flowering is
brief, it is unique in its beauty, and throughout the summer
and winter its close leafy growth is pleasing. It may be
grown from seed, but that method is a waste of time. The
best way to increase the plant is to tear it up in August or
September, and dib the pieces into a bed of rather poor soil
that has been well dug for the purpose. Showery weather
should be chosen for this operation, or water must be given,
and the plantation be kept shaded until rain occurs. The
variegated-leaved variety is a valuable rock and bedding
plant, scarcely so hardy as the common green-leaved plant,
and is likely to be destroyed in a severe winter or a damp
soil. The other species of arabis are not useful border plants,
but the variegated-leaved variety of *A. lucida* answers well
for edging beds on dry sandy soils, and makes a handsome
tuft on the rockery.

ARMERIA (Thrift).—The pretty plants of this family thrive
on rockeries and other similarly elevated positions, and on
dry sandy borders. They will also thrive on any good border
of the customary type, but a severe winter is likely to destroy
them when they stand on a cold damp soil. They may be
increased by division at any time during summer and autumn.
The best are, *A. alpina*, very dwarf, flowers reddish purple ;
A. cephalotes, a beautiful plant, with rosy crimson flowers ;
A. vulgaris, the Common Thrift of the cottage garden, of
which there are red, lilac, and white varieties.

ASTER (Michaelmas Daisy).—The plants of this family are

mostly tall gawky things, flowering abundantly in the later
autumn months. For the back rows of sunny borders they
are useful, and should, if possible, be planted away from
trees, though they bear partial shade with patience. They
are increased by division in spring. The best are *A. amellus*,
2 feet high, pale blue ; *A. elegans*, 2 feet, blue or purple ;
A. ericoides, 3 feet, white ; *A. nova angliæ*, 4 feet, reddish
purple ; *A. turbinellus*, 3 feet, purple blue.

ASTILBE.—Under this head we place the plant commonly
known as *Spiræa Japonica*, but which should be described as
Astilbe Japonica, for
it is not a spiræa,
but an alliance of
the saxifrage. It is
one of the loveliest
inhabitants of our
gardens, and very
much grown for sale
in the flower mar-
kets in early spring.
To grow this as a
border plant select
for it a damp shady
spot and a rich deep
soil. It will be found
perfectly hardy, and
far more likely to
suffer from the heat
of the sun in sum-
mer than from frost
in winter. Increase
by division when the
plant begins to grow
freely in spring. If
allowed to form large
tufts, it shows its
exquisitely beautiful
fern-like leaves and
feathery flowers to
great advantage.

ASTILBE JAPONICA.

AUBRIETIA.—This is sometimes called "Purple Alyssum,"
but it might with more propriety be called "Purple Arabis."

It is a first-rate plant for a sunny border and for a rockery.
Treat it as recommended for arabis. The best for the border
are *A. Campbelli* and *A. deltoidea.* Of the latter there are
several fine varieties, one of which has variegated leaves. It
is a lovely plant for the rockery, or to grow in a pot with
alpine plants.

AURICULA (Bear's-ear).—The great care bestowed upon the
valuable named varieties, that is to say, the florists' auriculas,
appears to place this plant at a disadvantage as one adapted
for the borders. Yet we have not a finer border plant, provided it has proper treatment. The common border, in which
all sorts of plants are grown, will suit them very well, as a
peep into almost any cottage garden will suffice to demon-
strate. But to enjoy them in an especial manner as border
flowers, prepare for them a selected spot, facing north, open
and breezy, and shaded from the mid-day sun in summer. There
need not be any elaborate preparation of the soil, but a deep,
well-drained, sandy loam is absolutely needful. If the plantation is to be a large one, it will be desirable to raise a stock
of plants from seed, and then the question arises, how to
obtain it? Shop seed of auriculas is, generally speaking, poor
stuff; but there may be somewhere a trader who can and will
part with a pinch worth sowing. As we are bound to give
direct advice, we counsel the amateur to purchase a few of the
named varieties of every class—selfs, white, grey, and green-
edged, and alpines. Grow these in frames the first season, and
save as much seed as possible. Sow the seed in pans filled
with fine sandy loam, and keep them in frames always moist,
until the plants appear, bearing in mind that you will have
to wait for them a considerable time. When the seedling
plants are as large as a bean, carefully transplant them into
pans or boxes, or into a bed in a frame, always giving plenty
of air, the use of the frame being advisable, because insuring
the plants more attention than they might obtain if planted
out in the open border in a very small state. When the stock
has increased sufficiently, plant out old and young in the
border, in the month of August, a foot apart, and leave them
to take care of themselves, remembering that the auricula is
one of the hardiest plants known, that drought is death to it,
that damp in winter is only a little less injurious. From the
time the first blooms of the seedling plants appear, a severe
selection must be made. Instantly, upon a bad flower

opening, pull out the plant and destroy it. By persevering in this course, and saving and sowing-seed every year, you will secure a fine " strain " of border auriculas, and if you can keep a border of about 150 feet length well filled with them, as we have done for many years, you will be able to prove, in the flowering season, that the auricula is one of the loveliest border flowers we possess. To perpetuate named varieties, divide the roots in July or August.

BEST FORTY-EIGHT AURICULAS.

Green Edge : *Booth's Freedom, Leigh's Colonel Taylor, Dickson's Duke of Wellington, Page's Champion, Hudson's Apollo, Oliver's Lovely Ann, Smith's Lycurgus, Cheetham's Lancashire Hero.*

Grey Edge : *Headly's George Lightbody, Turner's Ensign, Chapman's Maria, Turner's Competitor, Turner's Colonel Champneys, Reid's Miss Giddings, Fletcher's Ne Plus Ultra, Lightbody's Sir John Moore, Headly's Stapleford Hero.*

White Edge : *Campbell's Robert Burns, Heap's Smiling Beauty, Taylor's Glory, Smith's Ne Plus Ultra, Lightbody's Countess of Dunmore, Wild's Bright Phœbus.*

Selfs : *Spalding's Blackbird, Turner's Cheerfulness, Martin's Eclipse, Smith's Formosa, Lightbody's Meteor Flag, Martin's Mrs. Sturrock, Spalding's Metropolitan, Spalding's Miss Brightly, Turner's Negro, Chapman's Squire Smith, Headly's Royal Purple, Headly's Lord Clyde.*

Alpines : *Black Prince, Brilliant, Defiance, King of Crimsons, Constellation, Jessie, John Leech, Landseer, Minnie, Novelty, Venus, Wonderful.*

BELLIS.—The Daisy is a good though humble border flower. To grow it from seed is to make sure of a thousand worthless plants for one good one. There are in cultivation a number of beautiful named varieties, which should be purchased when in flower, if possible, and preferably if in pots. It is a sheer waste of time to plant any but the very best, and the best are cheap enough for the humblest amateur. They may be planted out at any time if taken proper care of, but the best time to plant, and also to take up and part for increase, is the month of August. In spring bedding the daisies play an important part in connection with anemones, arabis, wallflowers, and forget-me-nots.

CALTHA.—The Marsh Marigold (*C. palustris*) is not only one of the best things to plant beside a pond or stream, but a good border plant for a damp soil, and thrives in the shade. The *double-flowering* variety is the best: it may be propagated by division from October to March.

CAMASSIA (the Quamash).—This beautiful blue lily is a good companion plant to the agapanthus. It must have a damp, rich soil, and succeeds well in boggy peat. The flowers do not last long, but are charming in their brief day. Divide when the foliage begins to decay.

CAMPANULA (Bellflower).—The campanulas constitute a fine group of border flowers, which may be grown from seeds or divisions with the greatest ease, and thrive in almost any kind of soil if they but enjoy a moderate amount of sunshine. The only colours they offer us are blue, purple, and white, in various shades and degrees. They all flower in summer. The best are, *C. aggregata*, 2 feet, pale blue; *C. alpina*, 6 inches, dark blue; *C. macrantha*, 3 feet, deep blue; *C. glomerata*, 2 feet, purple, blue and white; *C. latifolia*, 5 feet, purple, a fine shrubbery plant for a poor soil, as it bears shade well; *C. persicifolia*, 2½ feet, blue. The beautiful *coronata* is a variety of *C. persicifolia*, and one of the finest of the whole group; *C. pumila*, a diminutive plant, flowering freely, blue and white; *C. carpatica*, dwarf, blue and white, a good bedding plant; *C. rotundifolia*, 1 foot, blue and white.

CARNATION.—See DIANTHUS.

CHRYSANTHEMUM.—This grand autumnal flower meets with but scant attention from the thousands of amateurs whose necessities and conveniences it appears exactly adapted to. We do occasionally see a few gay starry flowers in November in some entrance court, but rarely a border liberally furnished with the best varieties, and in such finished trim as Mr. Dale, of the Temple Gardens, presents them to public notice every year. To "do" them is easy enough, but the few attentions they require must be given them. They are increased by means of cuttings and division of the plants in spring, and it is well to provide a new stock every year, destroying the old stocks when a sufficient number of offsets or cuttings have been obtained from them. However much might be said about the cultivation of the chrysanthemum, all that it requires as a border flower may be summed up in fourteen words : Plant in a good soil and keep the plants securely staked from the first. All other matters are supplementary rather than necessary. To insure fine flowers, the soil should be well manured, and the plants freely watered, and the shoots should be reduced to six for each plant at the utmost, and the top flower-bud on each shoot should alone be allowed to remain after the buds have become fairly visible. In tying out, aim at forming a compact head, but allow space between the shoots for light and air ; for shade and confinement are most detrimental, though these are such excellent town plants. The Pompone varieties make magnificent beds, and are quite necessary for the border. When grown on sloping banks, the large-flowering kinds may be pegged down, to produce rich festoons and sheets of flowers. So, indeed, may the pompones ; but as the flowers of these are small, they are not so well adapted for surfacing, but they make most beautiful bushes. Very much being said in the books about " stopping" (that is, pinching out the shoots) it may be well here to say that when grown as a border plant, the chrysanthemum should never be stopped. The smallest plant put out in April will make shoots enough long before the time of flowering, and though stopping does increase their number, it causes the plants to flower later than they would do if not stopped, and that means a pretty certain loss of the flowers altogether, for frost may catch them before the flowers are out. The large-flowering kinds make good wall plants, and may be trained to low fences and dividing screens with advantage; for they are at least green all the summer, and in October and November make a splendid show of flowers.

BEST ONE HUNDRED CHRYSANTHEMUMS.

Incurved : *Abbé Passaglia, Beethoven, Beverley, Blonde Beauty, Bronze Jardin des Plantes, Dr. Brock, Duchess of Buckingham, Fingal, General Bainbrigge, General Hardinge, General Slade, Gloria Mundi, Golden Beverley, Golden Dr. Brock, Golden John Salter, Guernsey Nugget, Her Majesty, Isabella Bott, Jardin des Plantes, John Salter, Lady Hardinge, Lady Slade, Le Grand, Lord Derby, Miss Mary Morgan, Mrs. G. Rundle, Mrs. Brunlees, Mrs. Haliburton, Mrs. Sharp, Mr. Evans, Mr. W. H. Morgan, Pink Pearl, Prince Alfred, Princess Beatrice, Princess of Wales, Princess Teck, Rev. J. Dix, Yellow Perfection.*

Reflexed: *Alma, Cardinal Wiseman, Christine, Chevalier*

8

*Domage, Countess of Granville, Duc de Conegliano, Dr. Sharp,
Golden Christine, Golden Cluster, Julie Lagravère, Prince Albert,
Progne, Sam Slick, White Christine.*

Large Anemones: *Emperor, Empress, Fleur de Marie,
George Sand, Gluck, King of Anemones, Lady Margaret, Margaret of Norway, Mrs. Pethers, Prince of Anemones, Princess
Marguerite, Queen Margaret, Sunflower, Virginale.*

Pompones: *Adonis, Aigle d'Or, Andromeda, Aurore Boréale, Cedo Nulli, General Canrobert, Golden Aurore, Hélène,
Little Beauty, Madame Eugene Domage, Madge Wildfire, Miss
Julia, Mrs. Turner, Président Decaisne, Prince Kenna, Rose
d'Amour, Rose Trevenna, Salamon, White Trevenna.*

Japanese : *Bismarck, Dr. Masters, Emperor of China, G.
F. Wilson, Giant, Grandiflora, James Salter, Madame Godillot,
Nagasaki Violet, Prince Satsuma, Red Dragon, The Daimio,
The Mikado, The Sultan, Wizard.*

CHRYSOCOMA (Goldylocks).—The pretty *C. lynosyris* should
have a place in the front of the border, as one of the most
useful of " old things."

COLCHICUM (Meadow Saffron).—Plant in the front line
C. autumnale and its double varieties, *C. agrippina* and *C.
byzantium*, and leave them undisturbed for years. They are
really essential, as they flower in October and November,
when the border is likely to be dull.

CONVALLARIA—The Lily of the Valley (*C. majalis*) is a
most accommodating plant, and, generally speaking, needs
but to be planted in a shady spot and left alone, and it will
spread fast and far even to the extent of intruding on gravel
walks, and brick pavements. In cases where it refuses to
grow in this free natural manner, a small bed should be prepared in a shady spot, consisting of turfy loam from a fat
pasture, and in this bed the roots should be planted in the
autumn. There are some pretty varieties, the most beautiful of
them all is the *striped-leaved*, which, on account of its delicate
colouring in early spring, is usually grown in pots for decorating the conservatory. To obtain fine specimens, pot them
into nine-inch pots filled with a mixture of equal parts turfy
loam, rotten hotbed manure, leaf-mould, and silver sand, and
do not disturb them until they have quite filled the pots.

CORYDALIS (Larkspur Fumitory).—One of this tribe, *C.
lutea*, is one of our best garden friends, for it will soon form

a rich round tuft on the border, or spread over an old wall or ruin with rapidity to adorn the grey stone with brilliant sheets of yellow flowers all the summer long. *C. nobilis* is a fine plant adapted for rockwork, and requiring a deep gritty soil. *C. tuberosa* with dark purple flowers, and *C. t. albiflora* with pure white flowers, are two good border plants. The *C. cava* of the catalogues is properly *C. tuberosa.*

CROCUS.—This early-flowering cheerful old friend is quite appreciated, and we must not indulge in any moan on that score. Any soil will suit the crocus, but best of all a light rich sandy loam. The bulbs should be planted three inches deep in October. If kept out of the ground for any length of time they deteriorate seriously. A lot that we planted on the 1st of March with other bulbs in great part perished, and the few that lived did not flower. Yet in the first instance they were as fine bulbs as ever were seen.

DELPHINIUM (Perennial Larkspur).—This genus contributes to the border a splendid series of blue, purple, puce, and white flowers. They are mostly of medium growth, bearing par tial, but not heavy shade, though thriving more surely in the fullest sunshine; and all require a good deep rich mellow soil. Their fine qualities should command for them good cultivation. The first requisite is that they be carefully lifted every year in the month of November, and planted again after the places they occupy have been deeply stirred and liberally manured. They may be divided at the same time if desirable, but large clumps should first be secured. Another most important duty of the cultivator is to stake and tie the plants in good time, as the flower-stems rise in spring; and the third requisite is an abundant supply of water during seasons of drought in summer. The cultivator who cannot give them the attention required for the full development of their fine qualities may, nevertheless, do pretty well, for they are not fastidious plants, but they ought to be aided with stakes to make them safe against storms. They are not only good border plants, but grand bedders when carefully pegged down, so that the flower-stems rise about a foot or eighteen inches from the ground. The pegging down, however, is a nice business, and no one should employ delphiniums as bedding-plants until confident of the capacity to perform this operation without breaking the stems, or producing irregularity in the heights of the flowers. A peculiarly distinct display may be secured during the month of June by appropriating a large bed to delphiniums and scarlet geraniums. Some time in the autumn plant the bed with D. formosum or D. Hendersoni in lines eighteen inches apart, putting the plants nine inches to a foot apart in the rows. In the month of May, when the weather is settled and safe for summer bedders, plant between the delphiniums in close lines large old plants of scarlet geraniums that were pretty closely cut down in the early days of March. If the work is well done, the blue and the scarlet flowers will appear together, and produce a distinct and striking effect. As the delphiniums go out of flower, the bed will present scarlet flowers only. To raise delphiniums from seed is an extremely easy matter, but it requires much patience, for some of the sorts do not germinate for full twelve months after being sown. The seed should be sown as soon as ripe, and the pans should be kept in frames, and occasionally looked over, to remove weeds, which are sure to appear, and if allowed to grow will render useless all your labour. As all the members of this family are worth growing, the reader may select at random from a trade catalogue, but we select six which we consider most useful :—D. belladonna, 2½ feet, azure blue ; D. formosum, 3 feet, ultramarine blue ; D. Hendersoni, 3 feet, ultramarine blue ; D. Hermann Stenger, 4 feet, blue and rose ; D. magnificum, 4 feet, purplish or cobalt blue ; D. Wheeleri, 4 feet, bright blue. A few of the single kinds, and all the double ones, are sterile, and therefore can only be propagated by division or cuttings. To obtain the latter, cut down the plants in July, and in about a month afterwards they will bristle with tender shoots, which the cultivator must remove and make plants of.

DIANTHUS (The Pink).—Under this head we shall speak of the Carnation, Picotee, Pink, Sweet-William, and a few of their allies. The alpine pinks we shall have but little to do with, for they are not border flowers, but the popular members of the family are of the utmost importance for their beauty, fragrance, and comparatively docile habit under cultivation. All these plants require a good soil and a sunny situation, but a very fair display may be secured even if the ground is partly shaded and the soil not of the best. The florists pay so much attention to these plants, and bring them at last to such high perfection, that those who are unschooled in the " fancy " are apt to fancy that to grow a few good flowers is an almost super-

human undertaking. The truth is quite otherwise, as many
a cottager who has "blundered" into floriculture without
knowing anything of properties and exhibitions could attest
by the bonny pinks and carnations in his little garden. We
have had, and indeed still have, great clumps of cloves standing twelve years in the same borders, with hard woody stems
as thick as a child's wrist, and great twisted branches of the

CARNATION.

size of walking-sticks, and heads of grass covering a square
yard of ground, and these in the summer bearing hundreds
of grand flowers of the richest colour and most powerful
perfume. It is not in this way, however, that flowers of the
finest quality such as a florist would admire are produced.
One of the first requisites to success in the cultivation of
carnations, picotees, and pinks, is to acquire skill in propa-

gating them, in order to keep up the stock by means of young
plants.   This is the only important feature of the florist's
procedure that we need notice here, because our business is
simply to treat of them as border flowers.

Many readers of this work may be glad of information on
the essential characters of the three flowers we have now before

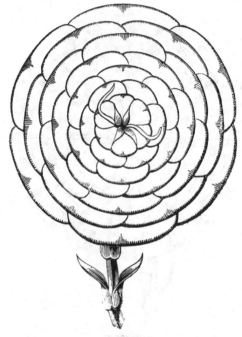

PICOTEE.

us.   It must be understood, then, that a Pink is heavily coloured
in the middle of every petal, this colouring constituting the
"lacing."   The Carnation is marked in flakes or stripes from
the base to the margin of every petal.   The Picotee is edged
with colour in marginal lines.   The Clove, or "girofler"
(Fr. *Giroflier*) of the old poets is a self-coloured carnation,

SHOW PICOTEE.
(Light red edged.)

possessing a powerful spicy perfume. For ordinary garden
purposes, the cultivation of carnations, picotees, and pinks is
the same, and therefore they may all be disposed of as one
plant, which will effect a saving of space, and enable the
beginner the more readily to master the first principles. We
shall begin by supposing the reader desirous of having a fair
show of all four classes of flowers, and our first advice is
that the purchase of plants should be made in the month of
September, and that
the whole of the
stock should be at
once planted out.
They may, indeed,
be planted in Octo-
ber and November,
and again in March
and April, but Sep-
tember is the best
time. In a well-pre-
pared soil and in an
ordinary good sea-
son they will require
but little attention
beyond being neatly
staked as the flower
stems rise ; but on
a hot dry soil, or in
an exceptionably dry
season it may be ad-
visable to give them
a good soaking of
soft water, or weak
liquid manure, once

PINK.

a week, from the middle of May to the end of August. It is
advisable, however, not to give water at all, if circumstances
favour their well-doing without it.

In keeping up the stock the two principal methods are by
layering and piping. Layering is performed from the middle
of July to the middle of August. One or two days before
commencing to layer give the plants a good soaking, unless
the weather happens to be showery. The operation of layering is performed as follows :—First strip off the lower leaves

of the shoots to be layered. Then take a shoot in the left
hand and bend it towards the stem of the plant with the fore-
finger, and with a small sharp knife in the right hand carefully cut the shoot half through, a little below the third joint
from the top, then turn the knife aside and slit the shoot
upwards about half an inch, so as to form a tongue. That
portion of the tongue which extends beyond the joint is to be
cut off and the shoot is ready for layering. Bend it down to
the ground and fix it with a hooked peg, keeping the tongue

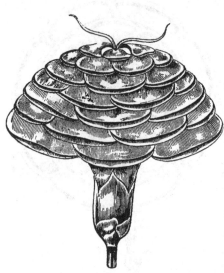

open with a pebble
or chip, and cover
the tongued por-
tion with one inch
of fine soil and the
operation is com-
pleted. The ap-
pearance of the
layer when pegged
down will be as
represented in the
figure. If dry
weather should fol-
low, the layers must
be watered, and
that is all the at-
tention they will
require until they
are rooted. Some
time in September
it will be well to
remove a little
earth from one or
two of the earliest

PINK.

layers to ascertain if they are well rooted. If they are,
they must be severed from the parent plant by cutting
through close to the joint at which they were layered, and
may be planted out at once, or potted singly in three-inch pots.
Our custom is to plant out a lot in clumps of three plants
each, six inches apart, in order to obtain a good show of bloom
the first year. In the autumn one or two plants are removed
to afford space for the full development of the one or two
remaining.

Piping is of less importance than layering, because it produces a less useful class of plants. The pipings are simply cuttings, prepared in a peculiar way. They are taken off in the last week of June or early in July, and consist of short jointed shoots, cut off close below the second or third joint, the bottom pair of leaves removed, and the base of the cutting split about a quarter of an inch. They may be

THE LAYER PEGGED DOWN.

struck under hand-glasses, or in Looker's Propagating-frames,

PIPING.—A, leaves to be removed; B, cut to the joint, and slit the base.

but the safest way is to plant them close together on a mild hotbed covered with six inches of light sandy soil. We have made thousands of useful plants by the rough method of the cottager, who grows everything he wants in the way of choice flowers by means of slips. The slips are made by pulling off the shoots; one or two of the lowest leaves are removed, and they are dibbed in thickly in some shady corner, and are as quickly as possible forgotten, unless the weather happens to be very dry, in which case they have a daily sprinkle of water to keep them cool and moist.

If the amateur grows any but the commonest sorts, the saving and sowing of seed will be an interesting and important business.   The finest varieties of carnation and picotee will yield but little seed ; indeed, we have found it a

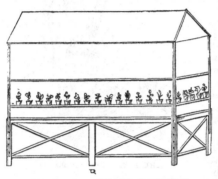

hard task to obtain a score of good pods from a hundred plants ; and when we had secured them, we would not have sold them for a guinea a grain. But how to obtain it, " There's the rub." The very commonest kinds will, for the most part, produce plenty of seed without any special care.

PROTECTING FRAME FOR CARNATIONS.

But those highly bred must have particular attention. In the first place, look to the semi-double flowers for the best supply. When you perceive that a seed-pod is swelling, pluck the petals one by one out of the calyx or cup, taking care at the same time not to injure the two horns (styles) in the centre. As the seed-vessel fills up, you may, with a pair of scissors, cut off the ends of the cup (calyx) all round, and make a slight incision down it, to prevent lodgment of wet. Towards the end of September the seed will be ripe, when it must be gathered and stored away.   It will be well to cut off the pods first and place them in bell-glasses in a sunny greenhouse, to ripen and shell out, as advised at page 99.   To raise seedling plants sow in pans in April, in good sandy soil, and cover the seed half an inch deep.   Keep only moderately moist, and have patience.   Above all things, do not push the seed forward in heat ; a cold frame is the proper place for the seed-pans. Grow the young plants on in beds of light soil, in a sheltered sunny spot, and plant them in the borders in August, or in a reserve bed in rows nine inches apart.   In planting, press the soil firmly to their roots, and finish with a good watering.

In places where hares and rabbits destroy carnations and pinks, they may be effectually protected by means of small covers made of common wood hoops and bramble branches, as represented in the figure.

Those species of dianthus which may be properly classed amongst alpine flowers are simply of no use at all in the herbaceous border, and therefore we shall pass them by. Our business is to find showy, ro- bust-habited plants that do not require the peculiar con- ditions which are essential to the well-doing and perhaps to the very life of the mountaineers. First amongst the most useful after carnations, picotees, and pinks, we must take the Sweet-William (*D. barbatus*), which is either a biennial or a perennial at the will of the cultivator. To praise this flower would be like " gilding refined gold," and so we abstain from eulogy, and say that seed may be sown in March or July. If sown early, under hand-glasses, or in a very gentle heat, the plants will bloom in the autumn of the same year ; if sown in July, they will not bloom until the following season. Our own preference is always for July sowing of seed newly ripe, and the planting out of the stock as soon as large enough where the plants are to bloom the following season. The sweet-william is remarkably hardy, and will endure severe winters on cold heavy soils, where car- nations would perish. There are some fine double varieties which never produce seed, and in every plantation single varie- ties occur which it is desirable to perpetuate. It is a quite easy matter to multiply these by cuttings, and the best way is to take for the purpose the blind shoots ; that is, the shoots that do not flower at the very time when the flowers are in perfection. The most simple cold-frame treatment is suffi- cient ; but it would be well to plant them out as soon as rooted, in order that, being well established, they may flower freely in the following season.

We have now almost done with Dianthus, but the section will be incomplete without a few more notes. *D. caryophyllus,* the Carnation or Clove in its unimproved or wild form, is a pretty little garden plant, with copious tufts of glaucous

grass-like leaves and small purple or white flowers. In like
manner *D. plumarius*, the wild pink, may be adopted as an
"interesting" plant to decorate rockeries and ruins, though
its white or purple fringed flowers must occupy one of the
lowest places in the ranks of floral beauty. *D. superbus*, the
superb pink, is a fine border and rock plant with pink, purple,
or white flowers, and deeply divided petals. *D. cruentus*, the
sanguineous pink, resembles the sweet-william, but is less
robust; on a dry sunny border or rockery it produces an
abundant display of its fine blood-crimson flowers. *D. dentosus*,
the toothed pink, is a dwarf tufted plant, with large purplish
flowers, that requires a warm dry border or sunny nook in a
rockery. *D. hybridus*, the mule pink, is of small growth and
extremely pretty, the flowers ranging from white to scarlet
in colour in the several varieties, of which there are about
half-a-dozen. The last-named section is admirably suited
for pot-culture, and are particularly valuable to supply cut
flowers in winter.

BEST THIRTY-SIX SHOW CARNATIONS.

Scarlet Bizarres : *Admiral Curzon* (Easom), *Captain
Thompson* (Puxley), *Dreadnought* (Daniels), *Duke of Welling-
ton* (Bragg), *Duke of Grafton* (Hooper), *Lord Napier* (Taylor),
Sir Joseph Paxton (Ely), *William Pitt* (Puxley).

Crimson Bizarres : *Anthony Dennis* (Wood), *Colonel
North* (Kirtland), *Eccentric Jack* (Wood), *Lord Goderich*
(Gill), *Lord Milton* (Ely), *Rifleman* (Wood), *The Lamplighter*
(Wood), *Warrior* (Slater).

Scarlet Flakes : *Annihilator* (Jackson), *Christopher Sly*
(May), *Illuminator* (Puxley), *Ivanhoe* (Chadwick), *John
Bayley* (Dodwell), *Mr. Battersby* (Gibbins), *Sportsman* (Had-
derley), *William Cowper* (Wood).

Purple Flakes : *Dr. Foster* (Foster), *Earl of Stamford*
(Elliott), *Florence Nightingale* (Seeley), *Mayor of Nottingham*
(Taylor), *Mayor of Oldham* (Hepworth), *Ne Plus Ultra*
(Hooper), *True Blue* (Taylor), *Squire Meynell* (Brabbon).

Rose Flakes: *John Keet* (Whitehead), *Mr. Martin* (Elkington), *Lovely Ann* (Ely), *Nymph* (Puxley) *Poor Tom* (May),
Queen Boadicea (Empsall), *Rose of Sharon* (Wilkinson), *Rosa-
belle* (Schofield).

Pink and Purple Bizarres : *Captivation* (Taylor), *Falcon-
bridge* (May), *Fanny* (Dodwell), *John o' Gaunt* (May), *Mas-*

terpiece (Schofield), *Purity* (Wood), *Sarah Payne* (Ward), *Shakespeare* (Puxley).

Heavy Red Edge : *Colonel Clerk* (Norman), *Countess of Wilton* (Holland), *Exhibition* (Elkington), *Favourita* (Kirtland), *John Smith* (Bonus), *Lord Valentine* (Kirtland), *Mrs. Brown* (Headly), *Mrs. Norman* (Norman).

Light Red Edge : *Ada Mary* (Smith), *Agnes* (Taylor), *Linda* (Fellowes), *Miss Holbeck* (Kirtland), *Miss Turner* (Taylor), *Mrs. Reynolds Hole, Countess Waldegrave* (Turner), *Wm. Summers* (Simmonite).

Heavy Purple Edge : *Admiration* (Turner), *Charmer* (Maltby), *Favourite* (Norman), *Lord Nelson* (Norman), *Mrs. Bayley* (Dodwell), *Mrs. Summers* (Simmonite), *Nimrod* (Fellowes), *Picco* (Jackson).

Heavy Rose Edge : *Aurora* (Smith), *Elise* (Kirtland), *Gem of Roses* (Gibbons), *Flower of the Day* (Norman), *Gipsy Bride* (Wood), *Pauline* (Fellowes), *Princess Alice* (Kirtland), *Scarlet Queen* (Wood).

Light-edge Rose : *Empress Eugénie* (Kirtland), *Lucy* (Taylor), *Maid of Clifton* (Taylor), *Miss Sewell* (Kirtland), *Miss Wood* (Wood), *Mrs. Fisher* (Taylor), *Purity* (Payne), *Rosy Circle* (Payne).

Light Purple Edge : *Amy Robsart* (Dodwell), *Bridesmaid* (Simmonite), *Ganymede* (Simmonite), *Lady Elcho* (Turner), *Mary* (Simmonite), *National* (Kirtland), *Princess of Wales* (Kirtland), *Rev. G. Jeans*.

Annie Chater, Beautiful, Beauty, Beauty of Bath, Bertram, Blondin, Charles Waterton, Christabel, Delicata, Device, Dr. Maclean, Edwin, Elcho, Emily, Eustace, Excellent, Excelsior, Flower of Eden, Invincible, John Ball, Lady Craven, Lady Clifton, Lizzie, Lord Herbert, Marion, Mildred, Mrs. Maclean, Mrs. Enfield, Perfection, Picturata, Prince Frederick William, Rev. G. Jeans, Sebastian, Sylph, The Pride of Colchester, Vesta.

DIELYTRA (China Fumitory).—There are about half-a-dozen species and varieties in cultivation, but only one, *D. specta-bilis,* a charming pink-flowered plant, and its white variety

D. s. alba, are worth growing. These two plants are alike in constitution, and may be spoken of as one for the purpose we have now in view. The hardiness of the Dielytra, is in a great measure determined by the nature of the soil in which it is grown. When planted in a dry sandy loam, it is rarely injured by the severest winter weather ; but, on the other hand, long-continued frost and snow will completely destroy the plants that grow in a deep, strong, damp loam. We have seen it standing five feet high, and broader across the head than a man could span, and then it was indeed indescribably beautiful. On our cold, heavy, damp soil it is comparatively useless, and we therefore grow it as a pot plant in the alpine house, and thus enjoy its elegant lively figure at the same time as the scillas, epimediums, drabas, and alpine primulas are in flower. The plant is easily multiplied by dividing the roots in autumn.

DIGITALIS (Foxglove).—The perennial species are second-rate things ; the Common Foxglove, *D. purpurea*, is a biennial, and must therefore be kept up by sowing seeds, unless, as commonly happens, after once obtaining a place in a garden, it maintains its position by means of self-sown seed. Where a considerable variety of herbaceous plants is required, the following may be planted—namely, *D. grandiflora*, 3 feet high, flowers yellow ; *D. ferruginea*, 3 feet, bronze coloured ; *D. ochroleuca*, 3 feet, pale yellow. They require a deep sandy loam, well drained, and it is well to put into the holes in which they are planted two or three whole bricks or large stones, so that the roots stand on a hard platform a foot or so below the surface.

DODECATHEON (American Cowslip).—Here is a charming little group of primulaceous plants, with flowers like those of a cyclamen. They require a rich, light, moist soil, and a shady situation, and should be taken up and divided in spring every three years. They may also be increased by seeds sown as soon as ripe in a cold frame. The best are *D. integrifolium*, flowers rosy crimson; *D. Jeffreyi*, very large leaves, and four-parted puce-coloured flowers; *D. meadia elegans*, rose and lilac ; *D. m. albiflorum*, white.

ERYTHRONIUM.—The Dog's-tooth Violet may be regarded as a companion to the American cowslip, though it belongs to the lilies, and not to the primulas. It will grow in a deep, light, mellow loam, or in peat or leaf-mould, or in heavy loam improved with a good admixture of old manure rotted to dust, and a considerable proportion of sand. We grow a few in the alpine house for the sake of their handsome spotted leaves as well as their charming flowers. Propagate by offsets as soon as the leaves have fairly perished. The best are *E. giganteum*, a splendid white-flowered kind ; *E. dens canis*, the common dog's-tooth violet, reddish purple; *E. Americana*, yellow.

ERYTHRONIUM GIGANTEUM.

FICARIA (Lesser Celandine).—This sweet little early-flowering British weed is most valuable for damp shady spots, where few other plants will grow, its bright green leaves and golden flowers being most welcome in the early spring. We have seen great patches in most unpromising spots in dark, damp, sour town gardens, and therefore it must have a place in this selection. All the varieties spread rapidly if the position suits them. There are four varieties : single yellow, double yellow, single white, and double white.

FRITILLARIA (Crown Imperial).—This noble plant should be fairly represented in every herbaceous border, and to grow it well it needs no skill at all; for the proper course of procedure is to leave it alone. Plant the bulbs in good deep loam in October. Take up and divide every three years. *F. imperialis* and its varieties, of which there are many, are alone worthy of general cultivation. The variegated leaved varieties are exceedingly beautiful. They make noble pot plants for the conservatory and for the plunging system.

FUNKIA.—A pretty group of liliaceous plants, with various and always handsome foliage. Any soil or situation will suit them, but rich sandy loam or peat is the most suitable, with partial shade. In a garden where snails abound they should

only be grown in pots in frames, for if the snails find them the
owner will lose them. The most distinct are *F. grandiflora*,
leaves pale green, flowers white; *F. ovata*, broad egg-shaped
leaves, and lilac-blue flowers; *F. Sieboldiana*, large ovate
glaucous leaves, and pale lilac flowers; *F. subcordata* (syn.
Japonica) *variegata*, an extremely beautiful plant, with pale
amber or cream-coloured leaves and white flowers.

GENTIANA (Gentian).—This is commonly regarded as a
troublesome genus, requiring some magical method of cultivation to insure a fair production of its notable deep blue
flowers. The magic consists for the most part in planting
properly in the first instance, and then leaving the plants
undisturbed for any length of time. We make a pilgrimage
occasionally to see a few great sheets of gentians bearing
thousands of flowers—a wonderful sight. The plants have
stood untouched for twenty years, and have travelled from
the border to the gravel walk, and compelled their owner
to make a new walk, to provide a way round them, this
being preferable to disturbing or destroying a single leaf or
root. It must be confessed, however, that the strictly
alpine species are fastidious and comparatively unmanageable,
and cannot be properly regarded as border plants. We shall
have nothing to say about the mountain gentians, and proceed
at once to say that *G. acaulis*, the Stemless Gentian, will grow
freely and flower finely in a deep, firm, moist, stony soil which
is neither clay nor sand. If a position is made for it, take out
at least a square yard of soil, one foot deep, and fill up with a
mixture of mellow turfy loam and large stones, and tread it
firm and plant. In the cottage gardens, where we occasionally
see it thriving gloriously, its well-doing is usually to be attri-
buted to its having obtained a soil to its liking, and having
been left alone to enjoy it. *G. asclepiadea* grows a foot and a
half high, has purplish flowers, and thrives on a deep rich
loam. There is a white variety: both are good border plants.
G. cruciata, with deep blue cross-shaped flowers, the plant
scarcely a foot high, will thrive in any good border. *G. lutea*,
the source of the druggist's " gentian root," is a handsome
plant, three feet high, with yellow flowers; it grows freely in a
deep rich moist loam. *G. saponaria* will thrive in any good
border; the plant rises a foot and a half, the flowers are blue
and barrel-shaped. *G. verna* is such a gem, that though really
fastidious, we must not omit it from this universal selection.

Find for it a *cold* and breezy situation in the border or rockery.

GENTIANA FORTUNEI.

The soil must be deep, rich, and cool, and so long as the plant is in the humour to grow, it must have constant supplies of

9

cold water. We began this selection with the intention of ignoring every troublesome and second-rate plant, and we break the rule here only because *Gentiana verna* is one of the loveliest plants in the world, and if it occasions a little trouble there will be found a few amongst our readers willing and glad to gratify its little whims and fancies. Any one with a soul big enough to poise on the point of a needle might feel a stirring of sentimentalism when beholding a great patch of the vernal gentian, quilted with flowers, in the month of April, and perhaps Campbell's song might suit the vein :—

> " I love you for lulling me back into dreams
> Of the blue Highland mountains and echoing streams —
> And of birchin glades breathing their balm,
> While the deer was seen glancing in sunshine remote,
> And the deep mellow crush of the wood-pigeon's note,
> Made music that sweetened the calm."

Cultivators of gentians may be thankful for a portrait of the tantalizing *G. Fortunei*, which, we are proud to say, was drawn from life. As we cannot keep the plant we cannot recommend it; but we shall hope for the day when the proper treatment of the plant shall be understood, when, no doubt, it will be found ready and willing to grow like a weed.

GERANIUM (Crane's-bill).—Very few of the hardy geraniums are worth a place in the garden, and those few have but to be planted and left alone and they will spread rapidly and thrive without care. The simplest way to multiply them is by division of the roots. The best are *G. pratense*, a handsome plant, with purplish-blue flowers ; *G. sanguineum*, well known, tufted, dark green leaves, and bright rosy purple flowers ; the variety *G. s. Lancastriensis* is better than the species ; *G. striatum* is extremely pretty, the flowers delicately pencilled, the leaves bright light green.

GLADIOLUS.—We must either say very much or very little under this head, and we elect to say the least possible. In warm, dry, sandy borders the finest kind of gladioli may be kept in the ground as hardy herbaceous plants ; but, generally speaking, they require to have special care in cultivation, and to be taken up in autumn and kept as dry bulbs through the winter. We have tried again and again to " acclimatize the named varieties of *G. ramosus* and *G. gandavensis* ;" in other words, we had left them out in the border, and have, except on a

few occasions, lost them wholly, so that in spring there were no
plants to be found on the sites where they bloomed the previous autumn. However, as hardy herbaceous plants, a few
species are available, and *G. cardinalis*, bright red, *G. insignis*,
orange red, and *G. segetum*, reddish purple, belong to this list,

GLADIOLUS.

because they were fine handsome plants, and will live through
the winter in any good well-drained border.

The garden varieties of the gladiolus have within the past
few years acquired immense popularity, the result in a great
measure of the immense improvements that have been effected
in the race by systematic cross breeding. We have now hundreds of named varieties, very many of them of stately habit
and remarkably sumptuous in colouring. The soil in which

these attain to fullest development is a rich gritty loam, containing a considerable store of vegetable matter, whether in the form of turf or leaf-mould. They thrive well in peat, and in any soil that is of a mellow texture and highly nutritive. The dry bulbs may be started in pots, in a pit or greenhouse, in February and March, and planted out in May; or they may be planted where they are to remain in the first instance, in the month of April. To be supplied with water in liberal measure, and have the support of neat stakes in due time, are the principal items in their management. They must be taken up as soon as the leaves begin to wither.

BEST FIFTY GLADIOLI.

Adolphe Brongniart, Belle Gabrielle, Brenchleyensis, Duc de Malakoff, Etendard, Eugène Scribe, Eurydice, Félicien David, Fénelon, Fulton, Galilée, Impératrice Eugénie, James Veitch, John Waterer, La Fiancée, Le Dante, Legouve, Lord Byron, Madame Dombrain, Madame Domage, Madame Furtado, Madame Vilmorin, Madame Adèle Souchet, Madame Basseville, Madame de Vatry, Madame Haquin, Madame Rabourdin, Mary Stuart, Maréchal Vaillant, Mathilde de Landevoisin, Meyerbeer, Michel Ange, Molière, Mozart, Napoleon III., Newton, Princess Clothilde, Princess Mary of Cambridge, Princess Mathilde, Princess of Wales, Rembrandt, Rev. M. J. Berkeley, Robert Fortune, Rossini, Semiramis, Sir J. Paxton, Sir W. Hooker, Schiller, Stuart Low, Thomas Methven, Thomas Moore.

GYPSOPHILA.—An extremely pretty genus, quite hardy, and peculiarly useful for bouquets, their tiny flowers, borne on slender stems, being like fairy filagree work amongst more showy flowers. Plant *G. dubia, G. paniculata, G. prostata,* and *G. saxifraga,* or any one of them, the second being the best if only one is required.

HELIANTHEMUM (Sun-rose).—These are pretty plants, but of quite secondary value. They are supposed to require hot, dry, sunny knolls, and certainly do well in such positions; but we find them quite hardy and prosperous on our heavy damp loam in a very cold climate. There are more than a score good varieties, alike in habit and differing in the colours of their flowers only. The following half dozen will please those who can find entertainment in their comparatively insignificant flowers:—*Croceum,* yellow; *Double Car-*

mine, carmine; *Rosy Gem*, rose; *Sudbury Gem*, crimson; *Singularity*, salmon yellow; *Miss Lake*,-primrose.

HELIANTHUS (Everlasting Sunflower).—These large-growing coarse plants are useful in large gardens and to make a blaze of yellow in rough half-wild places. *H. diffusus*, 4 feet, and *H. multiflorus*, 4 feet, are the best of them. Divide when needful.

HELLEBORUS (The Christmas Rose) is a grand plant, flowering from the end of the year to the middle of March, as the situation and the weather may determine. A heavy soil and a shady suits them all well, and it is of the utmost importance to leave them for many years undisturbed. In cold exposed places it is well to place hand-lights over the plants as soon as they begin to make new growth, in order to help the flowering, and the same practice may be resorted to for the production of an early bloom. *H. niger* is the best, the flowers are large, pure white, and resemble those of the water lily, though smaller. *H. olympicus* is worth growing, but none others are except by the insatiate searchers after uninteresting plants. Divide as needful in autumn, but the less disturbance the better.

HEMEROCALLIS (The Day Lily) is one of the best plants known for shady borders, and has but to be planted and left alone and it will do its duty. It is not a grand plant certainly, but its bright green sword-shaped leaves and bright ephemeral flowers are doubly valuable, because the worst situations will produce them in plenty. Increase by division, but allow the clumps to spread undisturbed for many years, if possible. *H. flava*, yellow; *H. fulva*, orange; *H. Kwanso*, double yellow, are the best. The variegated-leaved varieties are fine things for the border, or to grow in pots for the conservatory.

HEPATICA (Liver-leaf).—The lovely flowers of the hepaticas, produced in prodigal profusion in the earliest days of spring, outshine many of their companions of the garden borders, and best of all amongst a thousand suggest the fancy that the rainbows have changed to many coloured gems, and fallen in showers on the newly greened earth. So persistently do these beauties shrink from the hand of the careless cultivator, that when we meet with them in great flowery clumps, surpassing topaz, or sapphire, or ruby, or "orient pearl" in lustre, we know they have long been left to grow in their own sweet way, as those described by Milton in the happy garden :—

> " Flowers worthy of Paradise, which not nice art,
> In beds and curious knots, but nature born
> Poured forth profuse on hill, and dale, and plain,
> Both where the morning sun first warmly smote
> The open field, and where the unpierced shade
> Imbrown'd the noontide bowers."

It is easy enough to fail in the cultivation of hepaticas. Plant little mites in borders that are regularly dug and scratched, and altered and messed and muddled by that class of gardeners whose inborn faith it is that a tree exists only to be cut down and the prettiest weed to be pulled up; trust to this order of genius and you will never see any hepatica a second time. The amateur who has a fancy for perpetually transplanting, dividing, and improving, will never succeed with hepaticas, for the secret of success may be said to consist in first finding a proper place for them, and, secondly, in leaving them alone. Almost any soil will suit these lovely plants, but best of all a deep, rich, sandy loam—if stony all the better. Partial shade is better for them than the full sun, and a cold climate better than a warm one. When the clumps attain great size and rise up high above the ground, it will be advisable to lift and divide and plant again in soil deeply dug and refreshed with liberal manuring. The time for this operation is the autumn, when the growth of the season is quite matured. Where large masses occupy selected spots, it is advisable to spread over them in autumn a thin coat of dead leaves and short manure, through which the flowers will push in the following spring with increased vigour to make a more splendid show than would be possible without such aid. *H. angulosa* is a splendid species, with large sky-blue flowers. *H. triloba* is the best known, and there are about a dozen varieties of it, *all* of which are of equal value, so that to pick and choose amongst them would only be a waste of time.

HOLLYHOCK.—This grand landscape flower will never cease to be a favourite with the artists and the whole of that happy race who love the country, though the florists may solemnly assure us that it has fallen from its high estate. Fashion may vary the price of a thing, but it cannot enhance or depreciate the beauty of a single flower. To grow the

Hollyhock (*Althœa rosea*) in the garden border is a simple business enough, as may be learnt by observation. But to do it well, the soil should be deep and rich and damp, the situation open, and the climate gentle. It will grow well, however, on poor, dry soils, if aided with a good preparation in the first instance, and plentiful supplies of liquid manure afterwards. Sewage in a very weak state suits it admirably. Partial shade they bear well, but in deep shade they scarcely live. When standing on a damp soil, and especially in a cold locality, a severe winter is death to the hollyhock; but under moderately favourable circumstances, the plant is quite hardy, and if allowed to stand for a few years, acquires a buxom character, with its huge cluster of spikes, far to be preferred to the single spikes from young plants which content the florist. In making a plantation, secure pot plants of named varieties, the best of which are cheap enough for the humblest amateur. Plant in March or April, at three feet apart every way, arranging the plants, if possible, in accordance with their respective heights and colours. In a kindly season they will flower well if planted as late as May. They should be staked at the time of planting, or soon after, and be kept carefully and *loosely* tied as they advance; for if neglected, one small storm may tear the plantation to pieces. To propagate the named sorts, take cuttings from the base of the plant in August and pot them, and, if possible, promote quick rooting by placing them on a gentle bottom-heat. They must be repotted into separate pots, at least five inches in diameter, in October, and placed in a cold frame or greenhouse for the winter. Good seed will produce good plants, and therefore a stock may be got up quickly and with the most trifling cost by the amateur who can banish the word " trouble" from the garden vocabulary, and substitute " amusement " in place of it. If sown in February in a gentle heat, and grown on with careful regard to the fact that the plant is hardy and cannot well endure a strong heat, the seedlings may be planted out in May, and will bloom well the same season. Those who cannot manage them in this way had best sow the seed in July, and as soon as the plants are large enough to handle, plant them out in a bed of sandy soil in a frame, where they may remain until the time arrives for planting out.

BEST FIFTY HOLLYHOCKS.

Light : *Beauty of Milford, Cygnet, Carus Chater, The Queen, Royal White.*

Yellow, Orange, and Salmon : *Hercules, John Cowan, Junia, Leah, Mrs. Downie, Stanstead Rival, Orange Boven, W. Dean, Yellow Defiance, Excelsior, Gem of Yellows Improved, John Pow, Primrose Gem, Walden Queen.*

Crimson, Red, and Rose : *Captain Grant, Earl of Rosslyn, Fanny Chater, George Keith, Glory of Walden, Lady Dacres, Lady Vaux, Lady Rokeby, Mrs. Bruce Todd, Queen Victoria, Rev. E. Hawke, Royal Scarlet, Beauty of Walden, Crimson Royal, Garibaldi, Mrs. Hastie, Richard Dean, William Thomson.*

Lilac and Peach : *Countess Craven, Countess Russell, Lilac Perfection, Ne Plus Ultra, Willingham Defiance, Miss Barrett, Rose Celestial.*

Purple and Maroon : *Princess, Purple Emperor, Purple Prince, Othello, Black Knight, Lord Taunton, Purple Standard.*

HYACINTH.—This most valuable and early-flowering bulb is as well adapted for border culture as any plant in this list, though commonly regarded as a delicate thing that must be grown in pots with the aid of artificial heat. As "mixtures" of bulbs sorted in colours can be purchased at an extremely low rate of the seedsmen, and as a number of splendid named varieties may be obtained at a rate but little in excess of that charged for mixtures, and as, moreover, the simplest culture suffices to insure a brilliant display, there is every reason to favour a more extensive employment of the hyacinth in the British flower garden. A rich sandy mellow soil they must have, and if the weather is dry for some time when these flower-spikes are rising, water must be given abundantly. Plant the bulbs in October and November full six inches deep and six inches apart. If they push through extra early, owing to warm weather in December and January, spread over the bed a mulch of stable litter or cocoanut-fibre refuse to protect them from frost. This, however, will rarely be necessary, for they are not injured by frosts of ordinary intensity. Take up and store the sand as soon as the leaves decay. We have within view of the windows at the moment of writing this (May 2), a glorious display of hyacinths, tulips, and narcissi, which were only planted on the 1st of March previously.

That, of course, is an extreme case, but it shows that the cultivator has a range of full six months in which to purchase and
plant these bulbs.

THE BEST CHEAP HYACINTHS.

Single: *Amphion, Duchess of Richmond, Emmeline, L'Ami
du Cœur, Lord Wellington, Madame Rachel, Norma, Sultan's
Favourite, Grand Vainqueur, Grandeur à Merveille, Kroon
Princess, Baron Von Tuyll, Blue Mourant, Charles Dickens,
L'Ami du Cœur, Mimosa, Prince Albert, L'Unique, Alida Jacoba,
Heroine.*

Double: *Bouquet Royale, Grootvoorst, Princess Royal,
Waterloo, Anna Maria, La Tour d'Auvergne, Blocksberg, Lord
Wellington, Ophir d'Or.*

IBERIS (Candytuft).—The perennial candytufts rank with
arabis and alyssum in habit, season, and profusion of flowers.
Much might be said in their praise, but a few words will
suffice as to their cultivation. Any soil or situation, except it
be very damp or heavily shaded, will suit them, but they attain
to the finest development on a deep, dry, sandy loam, in an
open sunny situation, and are always more healthy and flori-
ferous when raised above the general level, as, for example,
on banks and rockeries. For masses of white flowers in the
spring garden more compact growing species are invaluable,
and as they may be grown with little trouble to a most perfect
state in pots, they answer admirably for plunging. They may
be raised from seed or cuttings, the latter being the better way.
The best time to take cuttings is when the young shoots of
the season are nearly full grown and are becoming firm. If
put in next the sides of pots filled with sandy soil, and shut up
in a cold frame, they will soon make plants. They should
pass the first winter in frames, and be planted out in the
ensuing spring. If seed can be obtained, sow as soon as ripe,
and grow the plants in frames until the following spring.
There are a few inferior varieties in cultivation which, of
course, are to be avoided. The best for massing, whether
planted out or in pots, is the true *I. sempervirens*, a compact
growing light green plant, producing an abundance of pure
white flowers. *I. Pruiti* has dark green leaves, the growth
dense, the flowers pure white, abundantly produced. *I. cori-
folia* is a valuable rock plant, and from its diminutive growth

adapted for forming a neat edging to beds in the spring garden. *I. Gibraltarica* is a very fine species. The plant is smooth, the growth tufted, the flowers white, in very large heads. In cold damp soils it is not hardy, but in dry positions in the southern counties is not harmed by the severest winter. *I. Tenoreana* resembles the last in growth, but is hairy, and the flowers soon change from white to lilac, or purplish red.

IRIS.—This is a great and grand family of garden plants, the real merits of which are at present known to but few, except the botanists, who, reversing the proper order of things, have obtained all the beauties of the family for their own enjoyment, while the world at large contents itself with the rubbish. A great tuft of common iris in a cottage garden is certainly no mean thing, but when we turn to the pretentious garden, the owner of which professes to have all the good things, we do not find the German iris, because it is " common," nor do we meet with such exquisitely beautiful plants as *I. reticulata*, *I. susiana*, or the pretty little *I. pumila*, or the variable and exquisitely painted " English " and " Spanish " iris. It has been truly said that amongst the species and varieties of iris occur flowers that rival the orchids in splendour of colouring, and may well stand in the stead of orchids in the garden where the costly exotics have not been domiciled. Fortunately the requirements of this family are few and of the simplest character, and admit, therefore, of being stated in very few words.

The family may be divided into two classes, the Rhizomatous and the Tuberous-rooted. In the first section the plants have fleshy, spreading, mat-like root-stocks or rhizomas; the second have tuberous roots, and for the better understanding of the distinction, may be termed bulbous-rooted.

The mat-rooted sorts claim attention first, as they are the most accommodating. They will thrive in any good garden soil, but when special attention is given them, the soil should be a deep, rich, moist loam. They thrive equally in sun and shade, but rarely attain to full development unless enjoying a few hours' sun from April to October. They are admirably adapted for planting in semi-wild places, and a few of them are especially valuable to adorn the margins of streams, and to fill up moist inlets about a lake or mere. Usually they produce plenty of seed, which should be sown as soon as ripe

on a bed of fine soil in a cold frame, or on a prepared plot on
a sheltered sunny border, the seed-bed to be covered with a
few branches of evergreens until the seedling plants appear.
Generally speaking, division of the root-stock in autumn will
be found a sufficiently rapid mode of propagating, as the
plants spread fast, and the smallest bit of root will "make
itself" in one season. They should be planted rather deep
according to the size of the roots, as they *grow upwards*, and
should be taken up every four or five years, and be planted
again deep enough to cover the crowns. When this is done,
the roots can be divided if desirable, and the ground ought
certainly to be deeply stirred and manured. The most
valuable species in this section is *I. Germanica*, the " blue
flag " of the cottage-garden. Of this there are many varieties, a few extremely beautiful, and many worthy a place in
the garden, for the sake of their singular markings and
curious shades of colour. Our fine British plant, the yellow
water iris, *I. pseud-acorus*, makes a grand mass of perennial
herbage, and a bonny show of yellow flowers in June, when
planted in a muddy inlet, or any odd bit of water waste. *I.
fœtidissima* is equally useful for positions a little less moist,
but likes to be near water. *I. graminea* is a good garden
iris, with flowers violet purple or yellow. *I. lutescens* is a
pretty little iris suited for a sunny bank or rockery, the
flowers are pale yellow. *I. pumila*, the dwarf, or Crimean
iris, is a charming plant for front lines and clumps in the
flower-garden, and worth growing in pots. There are about a
dozen varieties, of which the best are *cœrulea*, blue ; *versicolor*,
blue and white ; *atrocœrulea*, dark blue ; and *lutea*, yellow. *I.
susiana* is a grand plant for those who can grow it. The
requirements being a warm dry soil and a sheltered situation.
    The Tuberous or Bulbous-rooted kinds require a rich,
sandy, well-drained soil and shade from the mid-day sun in
summer. They all thrive in sandy peat, but there is no occasion to purchase peat for them in districts far removed from
peat-lands, because any good soil will be improved to suit
them by being well broken up, and plenty of old stable-manure,
leaf-mould, and sharp sand added to it. These kinds should
be planted only two or three inches deep, as they *grow downwards ;* and independent of the desirability of occasionally
dividing the roots, they must every three or four years be
lifted and planted again near the surface. *I. reticulata* is an

exquisitely beautiful little plant, with brilliant violet and orange-tinted flowers, that may be likened to violet velvet richly embroidered with gold lace ; *I. tuberosa*, the snake iris, is a curiosity not wanting in beauty ; *I. xiphioides*, the " English" iris, and *I. xiphium*, the " Spanish" iris, are charming things that increase rapidly by seeds, and vary in a delightful manner, the prevailing colour in both cases being what is called a " porcelain blue."

LATHYRUS (Everlasting Pea).—The showy plants of this family are well known for their rapid growth and splendid flowers. To cover low trellises, arbours, and the sunny parts of rockeries they are invaluable, and any good soil will suit them. They must have sun, or they can scarcely live. They make splendid displays if allowed to spread over a mound on the lawn, and indeed may be employed as bedding plants in any odd peculiar spots where colour is of more importance than neatness. The following are fine plants :—*L. grandiflorus* grows 5 feet high, flowers purple ; *L. latifolius*, 8 feet, purple ; *L. latifolius albus*, 8 feet, white, one of the very best for covering a mound ; *L. mutabilis*, purple, changing to red. The best mode of propagating is by division, but they produce plenty of seeds, which may be sown in pots, and the plants put out where they are to remain when large enough. We have never known the white everlasting pea come true from seeds, but it may be multiplied *ad infinitum* by cuttings.

LILIUM (The Lily).—The common white lily is, without question, the queen of the herbaceous border, and the very type of the interesting, handsome, hardy herbaceous plants we are searching for to arrange in this section. Amateurs who love collecting have here a grand field of operations, for the species and varieties are numerous, and, for the most part, equally beautiful and interesting. But for this selection, a few of the most distinct and showy kinds will suffice, and it will not be proper to multiply words in proportion to the importance of the subject, for those we shall select require but little cultivating, and are above the need of description and eulogy. All the liliums thrive in peat, and may, there-fore, be planted in beds of American plants, to show their fine flowers amongst the dense leafage of rhododendrons and azaleas. But they also thrive in deep, rich, mellow, moist loam, and in no case is it necessary to provide peat beds for them, or even to use peat when they are grown in pots.

A poor thin soil, a hot sandy or chalky soil, a peculiarly heavy
and wet clay soil are not suitable for lilies. In the improvement of the staple for them peat and leaf-mould are capital
agents, but well rotted stable manure is not less desirable; in
short, liliums are gross feeders, but a kindly mellow, well-
drained soil of some kind is indispensable for them. It is
commonly believed that lilies require shady aspects, but that
is a mistake. Some amount of shade they can endure without
injury, but the full sun is better for them if the soil is deep
and good to afford them a sustaining root-hold. The proper
time to plant is when the growth ceases, and the leaves die
down. Generally speaking, the months of August, September, and October constitute the season for planting lilies,
and the longer they are kept out of the ground (no matter
how carefully they may be packed), the worse will be their
condition when planted. The fact is, all soft fleshy bulbs
suffer by removal from the ground, and, therefore, when
liliums are transplanted, the site they are to occupy should
be prepared for them before they are lifted if possible, but if
they are to be planted again on the same spot, the work should
be done quickly, and the bulbs be, in the meanwhile, covered
with moist soil to protect them from the destructive influence
of the atmosphere. Generally speaking, they may all be multiplied rapidly by division when the leaves die down, and on
a pinch every scale of a bulb will make a plant if inserted
base downwards in a mixture of sand and fine peat, and
assisted for a time with greenhouse culture. But some of the
sorts ripen seeds in plenty, and if the seeds are sown as soon
as ripe in a bed or pan in a cold frame, a good stock of bulbs
may soon be secured. Some of the kinds produce "bulbits,"
or tiny bulbs on the flower-stems, and these, falling on the soil,
take root, and make an increase of stock that may prove a
perplexity to the cultivators. We have in our own garden a
collection of about a hundred species and varieties of liliums,
and some of the plantations are perfectly matted with young
brood, as if from seed sown broadcast, though all have been
produced from bulbits cast off by the flowering plants.

    We shall select eight sorts only. *L. auratum*, the grandest
of all lilies, is as hardy as the common white; at all events, it
has survived half a dozen winters on our cold wet soil in the
valley of the Lea, and is quite hardy in nurseries of Messrs.
Paul, of Cheshunt, and at the Hale Farm Nursery, Totten-

ham. *L. bulbiflorum*, the well-known orange lily, is indispensable. *L. candidum* is the most useful of all, though apt to become bare of leaves at the base of the flower-stem ere its season is over. To prevent this, lift and replant with a good dressing of manure in August, and give abundance of the water from the middle of May to the end of June. The variegated-leaved varieties make fine pot plants. *L. chalcedonicum*, the scarlet martegon lily, grows three feet, and produces a grand display of scarlet flowers. *L. longiflorum* cannot be left out, though on our cold soil it is nearly extinguished by a hard winter. It grows only two feet high, and

LILIUM LANCIFOLIUM.   LILIUM THUNBERGIANUM.

produces elegant funnel-shaped ivory-white flowers. *L. lancifolium*, in its several varieties, is quite hardy, but makes no show as a border plant; it is, in fact, lost amongst more showy species. It is, however, one of the best to plant in front of a rhododendron bed, as the dark green shrubs show up the elegant light-coloured flowers, and it is also a first-rate plant for pot culture. *L. Thunbergianum* is in the way of L. bulbiferum, but distinct enough, and there are several

fine varieties of it worth having, as, for example, *fulgens* and
venustum. Of *L. excelsum* a pæan of praise might be sung.
All we can do is to record that it grows four feet high, and
produces cream-coloured flowers. *L. tigrinum* completes the
select list; height 4 feet, flowers fiery salmon red. Several
of the popular kinds, such as the Turk's-cap and the Pyrenean,
we do not consider worthy of a place in a first-class border,

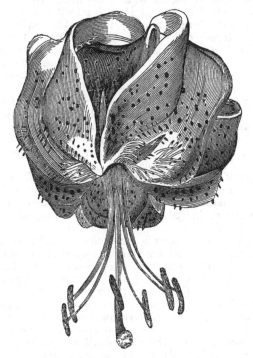

LILIUM LANCIFOLIUM.

and the little curious liliums that cost ten to twenty shillings
a bulb (we shrink as we call to mind the "heaps of money"
we have wasted on such things) are suitable chiefly for little
curious people. In peculiarly favourable spots, *L. lancifolium*
in variety, *L. giganteum*, *L. tenuifolium*, and *L. Leichtlinii*

may be added to enrich the collection, but they are not hardy enough for the universal garden.

LYCHNIS.—The British species that flowers in the hedgerows, almost outshine the best that belong to the garden. Still we must have a few, and grow them in moist light loam, in positions half shady. To multiply the best of them, cuttings of the flower-stems and division of the roots must be resorted to, but the least choice can be obtained from seeds. *L. alpina* is a charming little rock plant, with pink flowers; *L. chalcedonica*, 3 feet, flowers scarlet; the double variety better than the single; the white variety worthless. The double variety of *L. flos cuculi* is a charming plant, both white and red worth growing. *L. fulgens* is well known for its fine head of dazzling scarlet flowers. The double form of *L. viscaria* is also a first-rate border plant.

LYSIMACHIA.—The pretty "moneywort," or "Creeping Jenny," *L. nummularia*, is a capital plant for a shady, damp corner, and to plant on an old tree-stump, or on the edge of a vase. There is a golden-leaved variety good enough for a bedding plant. *L. thyrsiflora* and *L. verticulata* are good rustic plants for damp, shady borders.

LYTHRUM.—The lovely purple panicles of *L. salicaria*, rising from a watery nook or margin of a stream, have a peculiarly charming effect in autumn. The plant may, however, be grown in the border, if a moist, deep soil can be provided for it.

MECONOPSIS.—A near relation of the poppy, handsome and interesting. The species are few in number and peculiar in constitution. They will prosper best in light sandy loam and partial shade. *M. cambrica*, 1 foot, flowers pale yellow, is a fine plant. *M. Wallichi*, 3 feet, flowers pale blue, is a remarkably fine plant, difficult to grow, and probably a biennial. *M. Nepalensis*, 5 feet, flowers yellow, two or three inches in diameter. If tractable, this will prove one of the grandest of herbaceous plants. Those who dwell in the better climates of Britain, and have deep sandy or calcareous soils to deal with, should look after the species of Meconopsis as likely to prove of great value in the flower.

MIMULUS (Monkey-flower).—These are all lovers of a moist, rich soil, and do well in shady situations; provided they are not heavily overhung by trees, they increase fast enough for ordinary purposes by the spread of their roots, but may also be multiplied by cuttings. *M. cardinalis* is a fine plant, with scarlet flowers; *M. cupreus* is a little gem, with copper-coloured flowers; *M. luteus*, the yellow monkey flower, is a favourite of the cottage gardener; its varieties are numerous. *M. moschatus* is the Musk plant, which only needs to be planted in some shady nook, to run wild and become one of the best " weeds " of the garden.

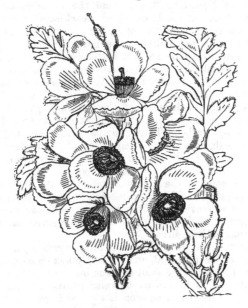

MECONOPSIS NEPALENSIS.

MUSCARI (Grape Hyacinth). — This beautiful group of plants is far too little known, and we trust many a reader of this note who has hitherto paid no attention to the grape hyacinths, will determine henceforth to be just to their merits. They may be grown in any ordinary garden soil, and will do equally well in sun or shade. It is desirable to lift and replant every three years. *M. botryoides* produces lovely spikes of sky-blue flowers, six to nine inches long; *M. comosum* produces purple flowers. A remarkable variety of this species is

10

pretty well known as the *Feather Hyacinth;* its flower-spikes are like marabout feathers. *M. racemosum* is a capital early-flowering kind, with deep purplish flowers.

MYOSOTIS (Forget-me-not).—Though few in number, and most humble in character, the garden is a blank that contains no forget-me-nots. Since "spring bedding" has been in vogue they have been in great request for their early display of myriads of light-blue flowers, and the introduction of a few newish and extremely beautiful varieties, which until of late were known to botanists only, has given quite a peculiar fillip to forget-me-not culture. As garden plants, they are short-lived, and perhaps have a better claim to a place in a chapter on annuals than in one on perennials. However, they *are* perennials; but those who would enjoy a perennial display of their charming flowers must propagate annually by seeds, or cuttings, or divisions of the root, and plant all out afresh in new, mellow, moist soil. *M. azorica* is certainly the best for the border, being hardy, comparatively robust, flowers at first reddish, afterwards deep blue. *M. dissitiflora* (syn. *M. montana*) is a most valuable species for early flowers, but "miffy," and therefore needing perpetual renewal. For growing in pots in the alpine house it is the best of all. *M. palustris,* the British forget-me-not, is too weedy for the border; but wherever there is a brook or half waste tract of marshy land, it should be planted, unless nature has taken care to locate it there already.

NARCISSUS.—As these can be grown anywhere, in sun or shade, in rich or poor soil, and multiply fast enough by the spread of their roots, we shall content ourselves with a selection simply. If it is desired to raise plants from seed, the proper course is to sow as soon as ripe, and grow in frames the first season. Then store away the dry bulbs until October, and plant where they are to flower. From the varieties of the Polyanthus Narciss, *N. tazetta,* we select as the best, *Sulphurine,* yellow and orange; *Glorious,* white and yellow; *Sir Isaac Newton,* gold yellow and orange; *Golden Beauty,* yellow and orange; *Grand Prince,* white and lemon yellow; *Grand Soleil,* deep yellow and orange. The double and single Jonquils, *N. jonquilla,* are eminently desirable, both for beauty and fragrance. The Poet's Narciss, *N. poeticus,* is delightfully fragrant, and its hardy and adaptive nature renders it suitable to plant in quantity in the shrubbery, and in the wilder parts of the grounds. The Daffodil, *N. pseudo-narcissus* is well known, and by no means to be despised, and is the more worthy of mention here because it offers a few splendid varieties, such as *Bicolor*, *Major*, and *Minor*, which are distinct in character, and admirably adapted to form interesting clumps in the shade of trees. *N. bulbocodium*, the "hoop petticoat," is an extremely pretty diminutive kind. *N. juncifolius*, the "rush-leaved" narciss, is an exquisitely beautiful miniature plant, adapted for the front of a rockery, and well worthy of pot-culture.

ŒNOTHERA (Evening Primrose).—The common evening primrose is but a poor representative of this fine family of showy, hardy, fragrant, interesting plants. The best herbaceous kinds will grow in almost any soil or situation, but thrive best in a light dry loam in full exposure to the sunshine. If, however, the soil is wet and cold they may be treated as biennials, for they make abundance of seed, and only need the most ordinary frame cultivation until the season arrives for planting them out. *Œ. Drummondi* is a neat downy plant, with bright yellow flowers; on a warm soil long-lived, on a cold soil it soon dies away, and must be kept up by means of seeds or cuttings. *Œ. Fraseri*, a neat plant, two feet high, producing abundance of bright yellow flowers. On our cold soil it stands well, and is one of our favourites for the plunging system. *Œ. macrocarpa* is a very showy and peculiar-looking plant, producing large yellow flowers. It requires a dry warm soil to stand its ground. *Œ. marginata*, 9 inches, flowers white and fragrant, stands well on any soil. *Œ. taraxacifolia* is in leafage like a dandelion; its large pure white, or pale yellow flowers are plentifully produced all the summer long, being in perfection soon after sunset.

PÆONIA (The Pæony Rose).—Gaudy, scentless, and short-lived are all the pæonies, yet no one who has seen a good plantation of the best sorts in flower would be in haste to exclude them from the select list of the handsomest and hardiest of herbaceous plants. When well-grown, every separate plant will form a mass of herbage equal in breadth to an ordinary flower-bed, or say, two or three yards across, and will produce forty or fifty flowers, each about the size of a man's head, borne on stout stems four or five feet high. They will live and flower in any soil, and in deep shade, and the worst place in a town garden will afford them a subsistence; but a deep rich moist loam, or a well-manured clay, and a full exposure to the sun, are the conditions that just suit them. A great clump of pæonies (of such sorts as we shall presently select) on a lawn near a pond, would make a sensational effect in the month of June, more especially if started with the help of a lot of manure, and kept going by the aid of an annual top-dressing put on in October. As any mite of a pæony root will soon make a plant, it is not needful to say much about propagating. The proper time to lift and plant large roots is from August to October, but pot-plants from nurseries should be put out in spring, and have abundance of water the first season.

The Herbaceous pæony is one of the hardiest plants in our gardens; not so its near relative, the Tree pæony, or Moutan, which is usually regarded as requiring the shelter of glass, and, under the best of circumstances, a most difficult plant to grow. The tree pæony is one of the many early-growing plants that suffer from keen east winds in a late spring; and hence, while it requires an open position, far away from walls and the shade of trees, it requires also the assistance of distant shelter, and a deep, rather dry, but exceedingly rich soil, and to be liberally aided with water all the summer.

BEST EIGHTEEN HERBACEOUS PÆONIES.

Alba mutabilis, Amabilis grandiflora, Antwerpiensis, Comte de Paris, Duchesse d'Orleans, Edulis superba, General Bertrand, Lilacina superba, Mathilde, Milbourni, Nivea plenissima, Pio Nono, Queen Victoria, Reine Hortense, Rosea plenissima superba, Tenuifolia flore-plena, Van Geert, Virginalis.

PANSY.—Exhibition pansies are grown in open beds of rich deep soil. The best time to plant is during September and October. Just before they come into bloom, they should have a top-dressing of rotten manure. (For border cultivation, see page 78.)

BEST FORTY SHOW PANSIES.

Selfs : *Arab, Cherub, Dr. R. Lee, Finale, George Keith, Imperial Prince, Locomotive, Miss Ramsay, Miss Muir, Ophire, Rev. H. H. Dombrain, Snowdrop, Virgo, W. Forbes.*

Yellow Ground: *Adam Scott, A. Whamond, A. Smith, Captain Sheriff, George Wemyss, George Wilson, John Baillie, J. B. Downie, John Downie, John Currie, Prince of Wales, Rev. J. Virtue, Thomas Martin, Victor, W. Martin.*

White Ground: *Cupid, Lady Lucy Dundas, Lavinia, Miss Addison, Miss M. Carnegie, Mrs. A. Buchanan, Mrs. H. Maxwell, Mrs. Galloway, Mrs. Hopkins, Princess of Wales, The Queen.*

BEST THIRTY FANCY PANSIES.

Avoca, Black Prince, Dewdrop, Eole, Earl of Rosslyn, Hugh W. Adair, Indigo, Lady Montgomery, Maccaroni, Magnificent, Miss M. Mather, Miss J. Kay, Mrs. Adair, Mrs. Laird, Mrs. R. Dean, Mrs. Shirley Hibberd, Mrs. H. Northcote, Magdalene Tweedie, Major Mackay, Miss C. Arbuthnot, Miss F. Hope, Pandora, Peter Campbell, Princess Mathilda, Rev. J. Robertson, Striped Queen, Sunrise, Wonderful, William Hay, William Baird.

PAPAVER (Poppy).—Only a few of these are worth mention. They must have plenty of room on a dry sunny border, and they will be gorgeous enough, but short-lived. *P. bracteatum*, 3 feet, flowers scarlet, is extravagantly showy. *P. alpinum*, a pretty little plant with yellow flowers, may have a dry sunny place in the front of the border. *P. pilosum*, 18 inches, flowers orange or brick-red, is also adapted for a dry sunny position.

PENTSTEMON.—Once more we light upon a splendid group of hardy plants, which are not well appreciated, because usually regarded as tender. It is true the garden varieties employed in bedding are apt to perish in winter on damp cold soils, but there are a few really hardy and most beautiful species and varieties to be found, and those that are not quite hardy may be kept on from seeds as soon as ripe, and the plants wintered in frames, and from cuttings made and kept in the same way as calceolarias, but as early in September as they can be obtained from the plants. Full exposure to sunshine is one of the first necessities of the pentstemon, and a deep, mellow, rich soil is scarcely less important. *P. barbatus*, 3 feet, with scarlet flowers, and *P. Torreyi*, a robust form of barbatus, are two of the best. *P. cobœa*, 3 feet, flowers variegated, needs to be kept on by means of cuttings, as it too often perishes in the winter. *P. Fendleri*, 1 foot, flowers light purple, quite hardy. *P. glaber*, 1 foot, deep blue. *P. procerus*, a trailing species with blue flowers, makes a fine tuft on a sunny ledge of a rockery. *P. speciosus*, 3 feet, flowers bright blue, hardy and handsome.

BEST THIRTY PENTSTEMONS.

Agnes Laing, Bons. Villageois, Arthur M'Hardy, George Avner, Arthur Sterry, Azurea elegans, Baroness Sempill, Candidate, Colin Bell, Harry King, James Forrest, James Rothschild, Grandis, John Pow, Lady Boswell, Magenta, Miss Carnegie, Miss Hay, Mrs. M. Binning, Miss Baillie, Mrs. A. Sterry, Novelty, Painted Lady, Purple Perfection, Purple King, Queen Victoria, Rev. C. P. Peach, Rosy Gem, Shirley Hibberd, Stanstead Rival, Sunrise, W. E. Gambleton.

PHLOX.—The immense number of varieties of phloxes in cultivation is evidence enough of the esteem in which they are held. They make sumptuous beds for autumnal display, and are unequalled for highly-dressed borders, and about the best of all known herbaceous plants to mix with roses, as they come into bloom as the roses give up for the season, and take our attention away from the jaded aspect of the queen of flowers. The garden phloxes, which have descended from *P. suffruticosa* and *P. pyramidalis*, are the phloxes *par excellence*. They are a most accommodating group of plants, for they will make a grand bloom on a poor soil, and last for years, becoming in time huge bushes that make a wonderful show in the late summer and autumn months. The way to grow them to perfection, however, is to renew the stock annually or biennially by means of cuttings, planting the newly-rooted pieces, in April, in rich deep loam well prepared for their reception some time previously, and giving water copiously, to promote vigorous growth until the plants come into flower. The stems are, of course, carefully staked as they rise, and the trusses are thinned to promote the production of large flowers. As to hardiness, the phloxes stand well on our heavy, moist land, where severe winters kill tritomas, hollyhocks, and pentstemons wholesale, therefore we may describe them as thoroughly hardy.

Amongst the more specific forms of phloxes, apart from the named varieties, the following deserve especial notice as first-rate hardy border plants :—*P. canadensis*, grows 9

inches to 1 foot high, the flowers are purplish lilac, produced
in abundance in April and May; a first-rate border plant.
P. frondosa, a dwarf spreading plant, with pink flowers in
April and May, very neat and pretty. *P. reptans* is a true
alpine plant in habit, that will grow anywhere; and if the
air is only moderately pure, thrives through the winter on
damp soils. It produces an abundance of pretty purplish
pink flowers in April. This is the *P. verna* of trade catalogues. *P. setacea* and *P. subulata* are small neat-growing
plants, with bristly leaves, wiry stems, and pink flowers.
They do not stand the winter well on damp soils, but are
good hardy plants, needing only a dry, open situation to
make a most welcome addition to the flora of the spring.

BEST TWENTY EARLY-FLOWERING PHLOXES.

*Adam Thomson, Duchess of Hamilton, Duchess of Suther-
land, George Goodall, Her Majesty, James Laing, James Neilson,
John Watson, Lady Abercromby, Lady Ross, Lewis Kidd, Mrs.
Austin, Mrs. Laing, Mrs. Hunter, Mrs. Murray, Mrs. Thom,
Princess of Wales, Robert Hannay, The Deacon, William Shand.*

BEST THIRTY-SIX LATE-FLOWERING PHLOXES.

*Aurantiaca superba, Adelina Patti, Comtesse de Chambord,
Comtesse de la Pannouse, Duke of Sutherland, Dr. Leroy, Etoile
de Neuilly, Géant des Batailles, John Laing, Liervalli, Madame
Domage, Madlle. Aubert Turenne, Madame Thibaut, Madame
Andry, Madame Barillet, Madame A. Verschaffelt, Madame
Marie Saison, Madame Rœmpler, Major Stewart, Mons. W.
Bull, Mons. Malet, Mons. Veitch, Mons. H. Low, Mons. Marin
Saison, Mons. C. Turner, Mons. Linden, Mons. G. Henderson,
Prémices du Bonheur, Professor Koch, Roi des Roses, Queen
Victoria, Souvenir des Fernes, Souvenir de Soultzmatt, Virgo
Marie, W. Blackwood.*

PINK.—See DIANTHUS, page 116.

POLYANTHUS.—As a border plant, *Primula elatior* is of the
easiest growth imaginable. Plant at any time, if the plants
are in pots; but if taken up from the open ground, the best
time is immediately after the fierce heat of summer has
begun to decline, and before autumnal frosts set in. A deep,
rich, moist loam, and a partially shaded position, are conditions favourable to this charming flower. The heat of

summer tries it much, unless it enjoys some amount of shade
and regular supplies of water. To obtain stock of named

LACED POLYANTHUS.

sorts, divide and replant in August. To raise seedling plants,
sow the seed in summer, as soon as ripe, or early in March,

and in either case grow the plants in a frame until large enough to plant out. Self-sown seedlings occur abundantly in the border where plants have flowered, and may be planted out in September or October. It is an important matter in managing the seed-pans not to allow the soil to become dry, for that is fatal to the germination of the seed. The varieties most prized are those with laced flowers, the ground colour being dark crimson, maroon, or black, and the lacing consisting of regular marginal bands of various shades of yellow or orange. The named varieties grown in pots for exhibition are of this class. The "giant" polyanthus are the most showy for the border and the parterre, being of all colours, and in many cases extremely beautiful. The following distinct varieties are particularly desirable :—*Double yellow, hose-in-hose, double white*, and *golden plover*.

POLYGONATUM (Solomon's Seal).—The common *P. multiflorum* will thrive in the shady border in the worst of soil, where scarcely any other plant can live, provided it is planted with a little care in the first instance, and then left undisturbed for years. The variegated-leaved variety, *P. m. fol. var.*, is exquisitely beautiful, and is much grown as a forced plant for exhibition. Easily increased by division when beginning to grow, in spring.

POTENTILLA (Cinquefoil).—A few of these claim notice on account of their showy flowers, but the genus is, as a whole, of comparatively small importance. The best, however, are but a short while gay, and all of them tend to untidiness in their mode of growth. Plant in the full sun ; any good soil will do. *P. atrosanguinea* is a fine plant, with deep crimson flowers. *P. Nepalensis* has scarlet or purplish-red flowers. Several fine hybrids of these have been obtained, the best of which are *Aurora plena, Grandiflora coccinea, Perfecta plena, Sudbury Gem, William Rollison, Aurantiaca*.

PRIMULA.—Under AURICULA, and POLYANTHUS, and PRIMROSE, three sections of this genus have been disposed of apart from the present selection of distinctive species. The alpine primulas are well adapted for border culture, if care be taken to plant them in damp shady spots, on mellow, gritty soil, elevated somewhat above the general level. Some few of them, however, must be grown under glass to be safe, and for such the alpine house or frame is the proper home. In any case, whether planted out or in pots, it is of the first

importance that the plants be sufficiently protected from
stagnant moisture by good drainage, and that they have
abundance of water in the growing season, and shade from
the fierce mid-day summer sun. *P. cortusoides* is one of the
best, and a true border plant. The leaves are heart-shaped,
light green, the flowers deep rose. A sandy loam suits it
well, and it is more likely to last out the winter on a rockery
than in a common border, because impatient of damp, but it
may be deluged with water all the summer to its advantage.
P. c. amœna is a variety of the last, with larger flowers, varying
in colour from delicate lilac and rosy red to the purest white.
Figures of two fine varieties of this primula were published
in the FLORAL WORLD of August, 1871. *P. denticulata*, with
toothed hairy leaves, and small lilac flowers, is a beauty to
grow in a gritty mixture of peat, loam, and sand on a well-
drained shady part of the rockery. *P. farinosa*, with leaves
densely powdered with meal, and lovely rosy lilac flowers,
requires the same treatment as the last; as does also *P. minima*,
a little gem with rosy flowers, which soon forms a precious
tuft on a rockery. *P. intermedia* comes near to the auricula
in character. It will do well in the border, if safe from
stagnant moisture in winter. *P. marginata* has a pretty tuft
of dusted leaves and pale lilac flowers. In constitution it is
like denticulata. These are all that we can venture to include
in the list, for other and equally beautiful species are so
impatient of the inevitable moisture of our winters, that
they must be grown in frames or alpine houses. Those
we have recommended may be increased by parting the
roots, and they will shed plenty of seed, which will germinate
without attention, and surround the parent plants with a
numerous progeny.

PRIMROSE.—Of the common primrose we shall say nothing.
Let those who love it not quickly expatriate themselves from
this land, or at least put down this book. The common
primrose can take care of itself; not so the uncommon
primroses, of which we shall strongly recommend a few as
absolutely indispensable to the border. It is the simple
truth that the very choice varieties of primrose are beautiful
beyond compare in their season, and an amateur who loves his
garden, and has none of these charming plants, is like the
philosopher in the sinking boat. (You know the story.) To
grow these precious pets, find a half shady spot on a good

border, and plant the sorts and leave them alone. That is, indeed, all you need do. But if your soil is arid, and your climate hot, you must keep the plants well fed all summer with weak liquid manure, or with soft water of some sort ; if from a ditch or pond all the better, as mayhap there will be something in it, for pure water is but poor food for plants. To obtain stock part and replant in August and September ; but, before doing so, give the plants a chance to spread, and make fine tufts, and show what they are. The following are so delightfully fresh, and distinct, and lovely, that, without any apology for the imperative mood, we plainly say you *must* have them :—*Single and double lilac, single and double purple, single and double rose, single and double white, single and double red, single and double yellow, single and double orange.* Here are twelve sorts in all, that may be purchased in good plants, at from sixpence to a shilling each, but as one of them, the single yellow, may be found on the nearest hedgerow bank, there is sixpence saved, which we beg you to spend on *another* plant of the double red, for when in flower the plant is as like a prize bouquet as a fanciful eye could desire.

PYRETHRUM (Feverfew).—The white-flowering "feverfew" is sufficiently well known ; as a bedding plant scarcely fit for gardens, but of the greatest value in parks and great rough places, where its tall weedy growth does not detract from its value as a plentiful producer of white flowers. Less popular, however, though more deserving of popularity, are the varieties of *P. roseum*, which offer us the most beautiful of all the flowers of May in the hardy herbaceous border. As, of course, many readers will not, for lack of actual knowledge, understand our estimate of their value, it may be well to say that the garden pyrethrum provides us at the dawn of summer with just such flowers as the asters furnish at the summer's close. Flowers quilled, anemone-centred, and variously formed and coloured, as asters are, with the advantage of hardiness, for they are as "hard as nails," and not even a damp soil hurts them seriously. To grow these fine plants to perfection, a good old garden soil is required, with plenty of manure, and liberal supplies of water in dry weather. The autumn is the best time to plant them ; but if they are purchased in pots, they may be planted at any time, except in the depth of winter. They are easily in-creased by division in August, or seed sown in a slight hotbed in February, or in a cold frame in April. As the best named varieties are cheap, the amateur will do better to obtain a good collection, and increase them by division, for seeds, however good the flowers from which they were obtained, invariably produce a large proportion of poor progeny.

The only pyrethrums, in addition to the varieties of P. roseum, that are worth a place in the border are the *double-flowering P. parthenium,* and the large single white *P. uliginosum.*

BEST EIGHTEEN PYRETHRUMS.

Album roseum, Candidum plenum, Emily Lemoine, Herrmann Stenger, Imperatrice Charlotte, Laciniatum plenum, Madlle. Bonamy, Michel Buckner, Monsieur Barral, Monsieur Calot, Mont Blanc, Nemesis, Paul Journu, Princess de Metternich, Purple Prince, Roseum bicolor, Rubrum plenum, Themisteri.

RANUNCULUS (Buttercup).—Although "bachelors' buttons" are old favourites, we really cannot recommend any of the

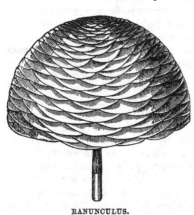

RANUNCULUS.

proper border ranunculuses, because of their coarse, weedy character, though we must confess a liking for the double varieties of *R. bulbosus* and *R. bullatus,* which the reader may elect, if he or she likes them. The florists' ranunculuses, descended from *R. asiatica,* scarcely belong to the border, but we dare not ignore such splendid hardy plants, and so we will endeavour to do justice to them in a short paragraph. The highest eulogy we can pronounce upon them is, that they are the most perfect of all florists' flowers in symmetry of form and perfection of colouring, and they are thoroughly hardy, well-behaved plants, adapted for any good border. The cul-

SHOW RANUNCULUS.

(White ground, red edges.)

tivation is the same as the anemone, but whereas that requires
a rather light soil, this requires a firm, well-holding loam.
They will, however, grow side by side in the same bed, in the
most friendly manner, in any good garden soil that is well
drained and prepared with proper care. Plant the tubers in
the first week of February, or as soon after as possible, in
drills two inches deep and five inches apart, the tubers five
inches apart in the drill. As soon as the plants push through
clear the ground of weeds, and tread it well between the rows.
Protect from frost as long as may be needful, and while dry
weather prevails, give water regularly until the flowering is
over, when watering must cease. Take up the roots when
the leaves have turned brown, dry them in a room or shed,
but not in the sun, and store away in bags or boxes.

BEST FORTY-EIGHT RANUNCULUS.

*Apollo, Ann Hathaway, Alexis, Beritola, Commodore Napier,
Coronation, Cedo Nulli, Camperdown, Delectus, Dr. Horner,
Eliza, Eva, Eupatoria, Festus, Gomer, Goldfinder, Herald, Hora-
tio, Indicator, Jenny Meldrum, Kilgour's Princess, Lord Gough,
Lord Berners, Marquis of Hereford, Melancthon, Miss Forbes,
Miriam, Mackenzie, Mrs. Guir, Melpomene, Miranda, Mrs.
Trahar, Naxara, Orange Brabançon, Orissa, Playfair, Pertinax,
Pelopidas, Princess Louisa, Prince Albert, Quilla Filla, Rose
Incomparable, Rubro magnificans, Sir W. Hoste, Sabina, Sophia,
Sir Philip Broke, Venus.*

RUDBECKIA.—A small group of showy asteraceous flowers,
which make a good appearance in sunny situations in the
undressed grounds in autumn, but are too coarse for a first-
rate border. The best are: *R. hirta*, 2 feet, yellow; *R.
laciniata*, 3 feet, deep yellow; *R. Newmanni*, 3 feet, yellow
and black; *R. subtomentosa*, 3 feet, yellow.

SAPONARIA (Soapwort).—A small group of alpine plants,
one of which, *S. ocymoides*, is employed for massing, on account
of its profuse production of lively pink flowers in the spring.
It requires a dry soil, and is admirably adapted for rockwork.
To increase the stock, take cuttings when the plants are
growing freely, or divide the roots in August. *S. officinalis*
is adapted for rough places, but not for the select border, as
it spreads its roots so fast as to become a nuisance.

SAXIFRAGA (Saxifrage).—This immensely large, various,

and beautiful family must be "broken up" for present consideration. A considerable number are true alpines, that
need peculiar treatment; others are fast-growing, and accommodating, tufty border plants that bear rough usage, and
almost refuse to die, though badly treated ; and all of them
are good rockery plants, that love partial shade and a deep
root-hold in gritty loam, where water cannot possibly stagnate.
They may all be increased by division of the roots, and by
seeds sown in a cold frame in spring. In selecting, we shall
begin with the large-leaved kinds, and recommend for the
border, but more especially for hillocks and rustic knolls, *S.
crassifolia*, which has broad and oval dark green leaves and
massive spikes of lilac flowers. *S. purpurascens* is a finer
plant, but scarce; the leaves are large and lustrous, the
flowers purple. *S. ciliata* is of smaller growth than the preceding; the leaves are hairy, the flowers white suffused with
pink. This large-leaved section is by some authors separated
from saxifraga under the generic distinction *Megasea*.

The best species of medium growth for borders are the
following : —*S. Andrewsii*, with tongue-shaped leaves and
conspicuous teeth, and flowers that somewhat resemble those
of the London Pride. *S. ceratophylla*, intense green in
leafage, and graceful panicles of snow-white flowers. *S. geum*
has kidney-shaped leaves, and beautiful white or pink flowers.
The *double* variety of *S. granulata* is a splendid border plant.
Lastly, this section would be incomplete if we omitted *S.
umbrosa*, the London Pride, one of the most accommodating
plants in the world, and one of the most elegant.

Amongst the smaller tufted-growing species, the best for
ordinary purposes are *S. cæspitosa*, which forms close cushions
of emerald green leafage ; the flowers are white ; a moist
position is one of its chief necessities. *S. hypnoides* is truly
moss-like in growth, and the best of the cushion-growing
kinds, as it will grow almost anywhere, if the situation is
moist and a little shaded.

When the cultivator has become accustomed to the ways
and wants of this interesting family, many more fine species
may be added to the collection, such as *S. oppositifolia*, *S.
cotyledon*, *S. hirsutus*, *S. diapensioides*, and *S. aizoon;* but
none of these are to be recommended for the mixed border.

SCILLA (Squill).—The best border plant in this genus is
S. nutans, the nodding squill, the *Hyacinthus non-scriptus* of

some botanists; the "blue-bell" hyacinth of the observant rustic. This plant will grow in any soil or situation, as its frequent appearance in splendid trim in damp and dark town gardens proves. It is certainly one of the best of wildings to introduce in wilderness walks and woodland scenes. There are several varieties, all of them good, comprising white, pink, flesh-coloured, and deep blue flowers. There are many more pretty squills in cultivation, and a few of them are employed for massing in the parterre. The best for general use are— *S. sibirica*, azure blue; *S. bifolia*, deep blue; and *S. b. candida*, a white-flowering variety of the last.

Sedum (Stonecrop).—From this hardy and useful genus the amateur may select almost at random, with the certainty of obtaining plants worth a place in any garden. Our old friend, *S. acre*, the common stonecrop, offers one of the best garments wherewith to clothe a sunny knoll, or to make a close mat-like edging on a somewhat dry soil. There is a remarkably beautiful variety of it, adapted for the spring garden, called *S. a. aureum*; its peculiarity is, that from Christmas to the end of May, the points of the shoots are of a bright gold yellow, producing almost as gay an

SEDUM SPURIUM.

effect as if the plant for nearly six months continuously was covered with flowers. *S. rhodiola*, the "roseroot," has a distinctive character which fits it for the border. *S. spurium* is a first-rate border and rock plant; the leaves are roundish and flat, fringed with transparent hairs, the flowers in loose corymbs of a bright rose colour. *S. telephium*, or orpine, is another good one, though common; when in flower, a great mass of it has a fine appearance. *S. fabaria* or *S. spectabilis*—the latter being the more correct name—is a large-leaved glaucous plant, growing freely in a bold tuft like a shrub, and producing fine heads of pale pink flowers in October. Neither

drought, nor damp, nor shade, nor frost ever harm this plant;
but it likes a sunny aspect and a good sandy loam. *S. Sieboldii*
is a most elegant creeping plant, with glaucous leaves set in
threes on arching whip-like stems; the flowers pink in September. In gardens where snails abound, it is simply impossible to keep this plant. It is a first-rate basket plant

for a sunny greenhouse, and the variegated-leaved variety is
even more handsome than the species.

SEMPERVIVUM (Houseleek).—There need not be much said
about the sempervivums, for they themselves will teach any
one how to grow them. They are at once interesting and
beautiful plants for sunny knolls, rockeries, roofs, walls, and

for edging flower-beds. It must not be supposed that they can live on nothing, though it is true they can get fat on short commons. In planting houseleek to adorn the roof of a shed, or the turret of an imaginary castle in an artificial ruin, something must be provided for it to live on, and there

SEMPERVIVUM ARACHNOIDES

can be nothing better than a mixture of fresh cowdung and good loam, smashed up together into a sort of putty. This can be laid in a heap where the plant is to be placed, and it will not slide, even from a rather steep slope. Insert the crowns with as much stem and roots as can be got, and fix them in

11

their places with bits of brick or stone pressed in beside them. The work is done, and you may rest from your labours for ten years at least. *S. arachnoideum* forms an elegant tuft, covered with white threads, as if enveloped in cobwebs; the flowers are purplish pink. *S. californicum* is the best for bedding, the leaves are dark green, tipped with brown. The offsets should be taken off in August, and potted in sandy soil, and wintered in a light dry pit for next year's use. *S. hirtum* is a close-growing hairy plant, producing myriads of white flowers, which the honey-bees will never leave while daylight lasts. *S. montanum*, producing purplish pink flowers, is another favourite with the bees. *S. tectorum* is the "houseleek" of the cottage roof, a good old homely plant that the heart will not willingly let die, though, for the matter of that, it is privileged with a thousand years' lease of its life, and will stoically defy a few kicks and scoffings :—

> Oh, such be life's journey, and such be our skill,
> To lose in its blessings the sense of its ill;
> Through sunshine and shower, may our progress be even,
> And our tears add a charm to the prospect of heaven!

SILENE (Catchfly).—These plants require a rich sandy loam and a pure air, and some amount of attention in the way of cultivation. As for multiplication, it is no vexation, for they produce seed in plenty, and cuttings of those with trailing stems can be struck in the summer with the greatest ease imaginable. *S. acaulis* grows in cushion-like tufts, the flowers reddish purple or pink. There is a white variety; both are good rockery plants. *S. alpestris* produces a lovely sheet of white flowers in May and June, and is well adapted for bedding purposes. A dry sandy soil, and the most free exposure to all the winds of heaven, are necessary to its well-doing. *S. fimbriata*, growing 2 feet high, and producing panicles of inflated white flowers, is a good border-plant. The *double variety* of *S. maritima* is a lovely plant for rockeries, if it can have a moist sandy soil. It is also a good bedding plant in a soil suited to its constitution. It grows about four inches high, and the flowers are pure white. *S. schafta* is one of the best for any and every purpose; a good border, rock, and bedding plant, nine inches high, producing reddish purple flowers in June and July.

SISYRINCHIUM.—A small group of interesting little irids

that require a light deep dry loam, or a good sandy peat. *S. anceps*, flowers bright blue, and *S. grandiflorum*, reddish purple, are the best for the border.

SMILACINA.—A sweet little gem is *S. bifolia*, requiring to be treated the same as lily of the valley, to which it is nearly related. It is invaluable for bouquets.

SOLIDAGO (Golden Rod).—These coarse-growing plants must not be ignored. *S. rigida* is the best, and quite worth having for its golden flowers in September. *S. altissima*, 6 feet high, is a good plant for the shrubbery.

SPIRÆA (Garland Flower).—All the herbaceous species are worth a place in the garden, and all require a deep moist loam, a few of them being amphibious plants that should always have their feet in cold water. *S. aruncus*, 4 feet, creamy white flowers, one of the best for margins of streams and moist woods. *S. filipendula*, a most elegant fern-like plant; the double variety is good enough for any border. *S. venusta* is a grand plant, with deep rose-coloured flowers. *S. palmata* is a very new and very grand herbaceous plant, as may be judged by the lively figure of it in the FLORAL WORLD of February, 1869, height 2 to 3 feet, flowers rich deep crimson. *S. ulmaria* is the proper "meadow-sweet," a delightfully fragrant rustic plant, with flowers like foam. "Everybody knows it" as an inhabitant of river-sides. The variety with yellow variegation is a good garden plant.

STATICE (Sea Lavender).—A few fine hardy plants are to be found in this genus, and we first of all recommend *S. latifolia* as a showy accommodating plant, the leafage and flowers of which will be prized for distinctiveness of character.

SYMPHITUM (Comfrey).—The plants of this family are showy, but coarse. A good loamy soil will suit them all. *S. bohemicum*, 2 feet, flowers brilliant crimson, a fine plant of its kind. *S. caucasicum*, 2 feet, purple, handsome. *S. officinalis* is the common comfrey, a coarse plant, worthy of attention for planting in damp woods, and by the side of streams, and also for its value as fodder. The variegated-leaved variety is one of the finest plants of its class for pot-culture.

THALICTRUM (Meadow-rue).—An unimportant genus, but any or all of which may be planted in capacious borders with the certainty of proving interesting. *T. aquilegifolium*, 4 feet, flowers creamy white; a good border plant. *T. anemonoides*, 1 foot, flowers white, graceful, and loving shade and moisture. *T. flavum*, 3 feet, flowers yellow ; a fine showy species, suited for the wilderness and shrubbery. *T. minus* is *almost* as elegant in leafage as the Maidenhair fern ; and a new variety, named *T. m. adiantifolia*, carries the similitude beyond the species.

TIGRIDIA (Tiger-flower).—These ephemeral flowers are so gorgeous in colouring, that we must bestow a few words upon their cultivation, with a view to direct the reader in the right path to an enjoyment of tigridias as they ought to be. To do justice to the flower, a bed should be prepared for it, consisting of good loam enriched with leaf-mould and hotbed manure, and the texture tempered with a sufficiency of sand to render it like potting compost. Plant the bed with bulbs of *T. pavonia*, four inches apart, in the last week of March ; keep the beds clear of weeds, give plenty of water in dry weather, and, when winter returns, lay a covering of litter on the bed, and fix it by means of a few withes and short stakes. The bulbs should not be disturbed more than once in seven years at least, and then they should be taken up, separated, and replanted in March in soil as well prepared as in the first instance. In a wet soil the roots must be planted annually, but they will never flower with the grandeur of those left for several years undisturbed. *T. conchiflora* is a remarkably fine plant, not so well suited to grow in beds as *T. pavonia*, but first-rate for pots.

TRADESCANTIA (Spider-wort). — The varieties of *T. virginica*, about a dozen in number, are good border plants, which should be left undisturbed many years. They like a rich soil, and endure patiently damp and shade. Their peculiar and elegant outlines fit them for nooks in the rockery, and to fill odd places where a mass of something distinct is needed. Divide in spring.

TRITELEIA.—A small genus of pretty white-flowered liliaceous plants. *T. uniflora* is the best, and will grow anywhere. Plant in autumn, and leave undisturbed three or four years. Though extremely elegant, the flowers are not good for cutting, in consequence of the garlic odour they emit.

TRITOMA (Torch Lily, or Red-hot-poker Plant).—This magnificent plant is one of the cheapest and most accommodating of the late-flowering lilies. *T. uvaria* is hardier than the hollyhock, and will thrive wherever the commonest lily can hold its ground. In common with most other good things, it grows most luxuriantly in a deep, rich, well-drained loam. In our damp, heavy soil in the valley of the Lea, it holds its ground well, and flowers most abundantly. *T. uvaria*, 3 feet, scarlet and orange, is indispensable. *T. u. glaucescens*, with extra long glaucous leaves, 4 feet, scarlet and orange, is more free to flower, and equally hardy. *T. u. grandiflora*, with very rigid scape, and flowers more decidedly scarlet than the others, is a grand plant, but the least hardy of the three. *T. media* is a good shrubbery species, flowering freely in the later months of the year.

TROLLIUS (Globe-flower). — A good border plant where the soil is heavy and moist, and bears

TRADESCANTIA VIRGINICA.

shade patiently. The best are *T. Asiaticus*, 1 foot, and *T. Europeus*, 2 feet; both have yellow flowers from May to July.

TULIP.—As a border flower the tulip has but one fault—it is short-lived. Of its splendour and variety we need say nothing—better is it we should make good use of what little space we can afford to say and prove that there is nothing in the catalogue of border flowers to equal the tulip in cheapness, adaptability to a variety of circumstances, hardiness, simplicity of management, and capability to make a liberal return for every reasonable outlay. Once become possessed of a variety worth growing, if the stock consists of but one bulb, and it not only need not be lost, but will be sure to increase yearly with the most trifling exercise of care and judgment on the part of the cultivator. Any ordinary good soil will grow tulips well, but the best possible soil for them is a well-drained, very rich and mellow sandy loam. Partial shade they bear well; indeed, it is the custom to put an awning over a bed of named late tulips, both to prolong the beauty of the flowers, preserve their true colours, and enhance the enjoyment of inspection, for a good bed of tulips is an exhibition in itself. For ordinary purposes, all the several classes and sorts of tulips may be treated in the same manner, and they will all flower superbly, and increase rapidly, and maintain their quality, though the circumstances they are subject to may not be such as a tulip-fancier would approve. Indeed, for the parterre and the mixed border no one needs expensive kinds; at the same time, those who have first acquired some experience in the management of the cheapest will be well prepared to plunge into the tulip fancy, if so minded, and they might do worse.

The early tulips are the most useful for massing, because they may be taken up in time to make the beds ready for geraniums and other summer bedders. They should be planted in October, four inches deep and six inches apart, and be taken up as soon as their leaves begin to wither, at the end of May. It is not necessary to wait until the leaves have quite died down; if they are but half dead, the bulbs may be lifted and laid aside, with a thin covering of earth, for a week, to ripen for storing. The late, or exhibition tulips, should be planted in November, and taken up in June, when the leaves begin to die down. It is no easy matter to kill tulips. We remember sending a valuable collection to the other side of the world, some twenty-five years ago. They were delayed in transit, and our calculations were upset. The result was, that nearly a year elapsed from the time they were taken up in England to the planting of the roots in the colony. Then when the boxes were opened, it was found that the bulbs had shrivelled away to dust, but every one had formed a cluster of tiny offsets to take its place, and from these offsets our friend soon obtained stocks of the several varieties that were sent out to him. In the autumn of 1869 we were so much occupied with big work, that the planting of our tulips was deferred, and deferred, until at last the 2nd of April, 1870, arrived, and

they were found much shrivelled and half grown in their
several drawers in the seed-rooms. On that day we planted
about three thousand bulbs on a piece of rough ground in the
kitchen garden. They had scarce a drop of rain (for it was a
season of drought), and were never watered nor weeded. At
the end of June they were taken up and stored away. In the
month of October following they were planted in the flower
garden, and at the time of writing this paragraph (May 2)
they are just go-
ing out of bloom,
having made a
glorious display.
Again, a lot of
early tulips, hya-
cinths, and narcis-
sus, bought in the
autumn of 1870,
were unavoidably
neglected until the
1st of March, 1871,
when they were all

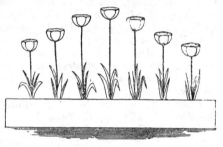

TULIP BED.

planted in the kitchen garden to " save their lives." On this
same 2nd of May they are all in perfection of flower, but a
great batch of crocuses, planted at the same time, have very
nearly perished.

The late or show tulips are well adapted for borders, in
which they can be left for several years ; but they are not
adapted for the parterre, because they cannot be cleared away
in proper time for the planting of the summer bedders which
should follow. When grown in proper florists' fashion, they
are planted in beds four feet wide, the sorts being arranged
so that they graduate in heights from the sides to the centre,
as in the subjoined figure. A bed of sixty rows of good
named show tulips—that is, 420 bulbs in all—may be obtained
for £20 ; and as for the early tulips, the prices of the very
best range from ten to thirty shillings per hundred.

A SELECTION OF TWO HUNDRED SHOW TULIPS.

The following is a list of 200 cheap first-class sorts, which
every beginner should possess, as they stand in the foremost
rank at all our great exhibitions :—

BIZARRES.—First Row : *Albion, Dr. Horner, Goldham's*

Fortunius, Golden Fleece, King of Tulips, Marshal Soult, Osiris, Roi de Navarre, Groom's Rubini, Sir Edward Codrington, Lawrence's Solon, Lawrence's Selim, Stein's Napier, Telemachus, Clarke's Ulysses.——Second Row : *Ariadne, Apollo, Bizard Le Kaine, Coronation, Charbonnier Noir, Captain White, Darius, Lawrence's Glencoe, Gloria Mundi, Lawrence's Ostade, Optimus, Lyde's Oddity, Pilot, Lawrence's Peacock, Strong's Titian, William IV.*——Third Row : *Carter's Leopold, Charles X., Captain Sleigh, Delaforce's King, Lawrence's Fabius, Lord Strathmore, Lord John Russell, Magnum Bonum, Milton, Ophir, Polyphemus* (feathered), *Polyphemus* (flamed), *Prince of the Netherlands, Strong's Hero, Salamander, Walker's King.*——Fourth Row : *Dickson's Duke of Devonshire, Lawrence's Donzelli, Emperor of Austria, Lord Collingwood, Proteus, Sharp's Victory* (alias *Sultan*), *Lawrence's Sheet Anchor, Warsaw.*

BYBLŒMENS. — First Row : *Bienfait, Chellaston Beauty, Euclid, Gloria alborum, La Belle Narine, Parmigiana, Goldham's Prince, Queen of the North, Strong's Claude, Gibbon's Purple Perfection.*——Second Row : *Lawrence's Friend* (alias *Addison*), *Brown's Wallace, Bijou des Amateurs, Blomart, Cleopatra, Countess of Harrington, Lawrence's Diogenes, Euterpe, Gibbon's Enchantress, Grand Monarque, Irlandois, Ivanhoe, Joseph Strutt, Lalla Rookh, Lewald, La Virginité, Lawrence's Lord Stanley, La Joie, La Latière, Malibran, Maid of Orleans, Mentor, Gibbon's Purple Perfection, Penelope, Prince Charles, Reid's Prince Albert, Wilmer's Queen Victoria, Queen Charlotte, Rubens, Smith's Wellington, Superb et Noir, Victoria Regina, Violet Blondeau, Violet Rougeâtre, Winifred, Zoe.*—— Third Row : *Acapulca* (alias *Roi de Siam*), *Gibbon's Britannia, Black Baguet, Cincinnatus, Colossus, Desdemona, Duc de Bordeaux, Duc de Bouffleurs, Gibbon's Elegans, Franciscus Primus, Grotius, Grand Sultan, Holme's King, Lawrence's Lady Errol, Lawrence's Lord Hawkesbury, Michael Angelo, Miss Porter, Princess Charlotte's Cenotaph, Princess Royal, Lawrence's Patty, Lawrence's Priam, Tintoret.*——Fourth Row : *Ambassador, Alexander Magnus, Lawrence's Camarine, Captain Lampson, Commodus, Lawrence's Elthiron, Louis XVI., Saint Paul, Thalia, Violet Quarto, General Barnovelde, Hugobert, Lelot Sovereign, Lilias' Grand Vase, Pass Salvator Rosa, Carter's Regulator, Wood's Rembrandt, Sir H. Pottinger, Gibbon's Surpass Le Grand.*

ROSES.—First Row: *Scarnell's Bijou, Cerise Blanche, Catalina, Fleur des Dames, Kate Connor, Madge Wildfire, Rose, Juliana, Lady Diana Boyle, Lachesis, Lady Wildair* (flamed), *Ondine* (feathered), *Groom's Persiana, Rose mignon.*——Second Row: *Aspasia, Andromeda, Cerise à Bella Forme, Comet, Lawrence's Cymba, Duchess of Newcastle, Groom's Duchess of Sutherland, Dutch Ponceau, Slater's Fairy Queen, Goldham's Maria, Lawrence's Lady Waldegrave, Clark's Lavinia, Mary Lamb, Mason's Matilda, Perle Brillant, Perle d'Orient, Rose Imogene, Triumph Royal, Strong's Duchess of Kent, Lawrence's Emily, Willison's Juliet, La Belle Nanette, Ponceau-très-blanc.*——Third Row: *Lawrence's Aglaia, Anastasia, Claudiana, Lawrence's Duchess of Clarence, Fanny Cerito, Lord Byron, Rose Camuse, Rose Brilliant, Rose Galatea, Lawrence's Mary Anne, Rose Cordelia, Rose Walworth, Thalestris, Hayward's Magnificent, Vicar of Radford.*——Fourth Row: *Lawrence's Clarissima, Comte de Vergennes, Lawrence's Emily, Madame Vestris, Mountain Sylph, Midland Beauty, Prince William IV., Rosa Blanca.*

BEST THIRTY BEDDING TULIPS.

Red: *Cramoisie, Vermilion Brilliant, Couleur Cardinal, Monument, Feu d'Anvers, Zongloed, Van Thol.*

Yellow: *Marquis de Nesselrode, Yellow Prince, Yellow Tournesol, Yellow Rose, Grenadier, Yellow Pottebakker.*

White: *Alida, White Pottebakker, Jagt van Delft, Luna, Nonsuit.*

Various: *Roi Pepin,* white and crimson; *Duc d'Aremberg,* crimson and gold; *Florida,* deep mauve; *Keizerkroon,* crimson and gold; *Thomas Moore,* yellow and buff; *Van der Neer,* puce; *Proserpine,* crimson; *Bonaparte,* chocolate.

Double: The best doubles for a group are *La Candeur, Rex Rubrorum, Tournesol, Yellow Rose.*

VERONICA (Speedwell).—The shrubby veronicas are not quite hardy, but must have place here on account of their massive character and showy flowers. They answer to plant against dwarf walls, and in peculiarly sheltered, sunny, well-drained positions. They may in cold climates, and on damp soils, survive several winters in succession, and at last disappear suddenly before the assaults of cold and wet. Any ordinary good soil will suffice to sustain them well, and the poorer and drier the soil, the hardier will be the plants. They may be most easily increased by cuttings of young shoots in summer; and being most easy plants to manage, may be grown in quantity in pots for the conservatory, and to make pleasing masses in the garden in the autumnal months by plunging the plants when in full bloom in a suitable border. The best of them are *V. Andersonii*, blue and white; the *variegated leaved* variety of the same is much used in bedding, and makes a fine conservatory plant; *V. decussata*, blue; *Gloire de Lyon*, crimson; and *multiflora*, violet and white. The herbaceous veronicas are an inferior lot of plants; but *V. amethystina*, and *V. spicata*, are worth a place in the border and require only the most ordinary treatment.

VINCA (Periwinkle).—The fast-growing, shade-loving, most accommodating and beautiful hardy vincas are things almost unknown to the majority of amateur gardeners. There is no end to the uses they are adapted for; but to clothe banks and half-waste spots under trees, and to fill up nooks where scarcely any other plant will grow, they are invaluable. The collector of good herbaceous plants should make it a point to secure all the sorts, and plant them somewhere in view of the possibility of needing some day to propagate a stock for some particular purpose. The young shoots may be struck in summer under hand-glasses, or they may be pegged down to root around the parent ready for removal next season. All the sorts are good, and they number in all about a dozen. · *V. reticulata* is a bold showy plant, with leaves rich green, and prettily pencilled; *V. major fol. var.* makes a good edging to flower beds, and being quite hardy is a capital poor man's substitute for variegated geraniums ; *V. minor* forms a neat little tuft, which in spring produces more blue flowers than any other kind.

VIOLA (The Violet).—Here again we are tempted to say much, but intend to say little. In our deep heavy land, violets of every kind grow with astonishing vigour, and flower with extravagant profusion without any care at all. We might be tempted, therefore, to advise leaving the plant to take its chance as a weed in the garden, did we not happen to know that in many cases it must have systematic treatment, or it never justifies its occupation of the soil. Happily, we can sum up the case in a few words. In the first place, all kinds of violets that are worth growing require a good rich moist soil and a shady situation. It is in the mellow product of rotted leaves, and the warmth and shade of the wood, that nature brings forth violets to perfume the breath of the spring. In preparing a soil for violets, use leaf-mould and very rotten hotbed manure freely. If the soil is strong, but unkind, dig in a great quantity of charrings from a smother. Having secured a proper soil, the next most important matter is to raise a stock of plants every year. The simplest mode of doing this is to take up a lot of old plants and tear them up in May and plant them in fresh soil. A far better way is, about the middle of April, to spread amongst the plants a mixture of leaf-mould and rotten manure, working it in by means of a broom or the hand, when the plants are quite dry. After this water the bed frequently with a waterpot fitted with a fine rose, to keep the surface-soil moist. In about twenty days there will be newly-rooted runners all over the bed holding to the tempting stuff with which the plants were top-dressed. Now dig them all up, remove the strongest of the young newly-rooted runners, and plant them in a well-prepared bed, and throw all the rest away. Keep the plantation well watered during dry weather until the end of August, after which water need not be given. In due time you will have plenty of violets. If turf pits can be spared it is a good plan to plant in them a lot of the earliest and strongest runners, and then by putting on the lights as soon as the chilly nights of autumn return, the plants will bloom three months in advance of those in the open ground. There are many varieties of sweet violets in cultivation, and some of them are good, such as *The Czar*, and the *Giant;* but for out-door growth there is nothing to surpass the *Russian*, and for frame and greenhouse culture the *Neapolitan*. The so-called red violets are ill-looking, and scarcely sweet; the white-flowered are elegant and delightfully fragrant.

The border violas are mostly American, and scentless. The best are : *V. cornuta*, pale blue; *V. lutea*, bright yellow ; *V. palmata*, purple; *V. pedata*, dark blue ; *V. tricolor*, the common "Heartsease," for cultivation of which, see PANSY, page 78.

WALLFLOWER.—This is commonly classed with annuals, and, as such, is one of the most useful of our hardy plants. We place it here, because the real wallflower, *Cheiranthus cheiri*, and all its relatives, are true perennials, and may be grown from year to year, until they acquire the character of miniature trees, four or five feet (or more) in height. Though capable of existing almost anywhere, the common wallflower is scarcely a thriving plant in shady positions and on cold, wet soils. Warmth and dryness are important conditions of its well-doing, and it will attain complete development on a wall or buttress, where it has but a mere spoonful of dust to root in, while on a rich, heavy soil, it will progress but slowly, and will surely die in a cold, wet winter. A light, rich, and well-drained sunny border will suit all the plants in this section; damp is always death to them, but they scarcely suffer if required to grow in partial shade. It is so easy to get up a stock of wallflowers from seed, that we shall be content to advise that, if a succession of flowers is desired, three sowings should be made—in April, May, and June; and the open border is the best possible seed-bed. To plant them out as soon as they are large enough to handle is an important matter; for if they remain crowded in the seed-bed, they become attenuated and comparatively worthless. Those wallflowers will bloom the best that have been long standing on the same spot; and, when removal is necessary, it should be performed in dull, showery weather. We will suppose, now, that you are enjoying the cheerful appearance and delightful odour of a mass of wallflowers, and you note amongst them a few with particularly fine flowers. If you wish to keep those varieties for any special purpose—say for spring bedding—the simplest and safest course will be to take from them as many cuttings as possible, and strike them under hand-glasses or on a mild hotbed, and the stock is secured. When the plants are in bloom is the proper time to make the cuttings; and the blind shoots at the base of the plant—that is to say, the small green shoots that have not flowered, are those which should be removed to be made into plants. There are in cultivation a few peculiar " strains" of wallflowers—one in particular, a dwarf bushy plant, with flowers of the clearest yellow. There is much difficulty in obtaining seeds of these highly-valued varieties, but having once secured a pinch of true seed, or a few plants of the right sort, the cultivator never need lose any of them again, for he has two strings to his bow—he may save seed and strike cuttings; and though the first may sport away from the proper type, the second will not, but will reproduce exactly the characteristics of the parent plants. The *double-flowering* varieties can only be perpetuated by cuttings, and those who purchase seed " warranted " to produce double flowers are sure to be disappointed. The old *double yellow* is a grand plant when well grown, both for the conservatory and the open border. The sort of border that suits wallflowers best is one adjoining the wall of a greenhouse, and the soil should consist of equal parts good sandy loam, and broken bricks, and old mortar, two feet in depth, resting on a dry subsoil. In such a border the double wallflowers will live for many years, and become handsome trees. Any aspect will suit them, and if four walls with four several aspects can be planted, there will be a succession of flowers from the turn of the year on the south side, to quite midsummer on the north.

The alpine wallflower, *C. alpinus*, growing 6 to 9 inches high, with flowers of the brightest yellow, is a valuable plant for a dry border or rockery, and it bears shade well. Marshall's wallflower, *C. Marshalli*, is extremely neat in growth, and remarkably showy when in flower ; it grows one foot high, and the flowers are of a deep orange colour. Both these can be grown from seeds or cuttings, and where they are employed in spring bedding, it is important to make sure of them by means of cuttings ; for they cannot be depended upon to produce good seed in plenty : that, indeed, depends very much upon the peculiarities of the soil and the season. Those who are anxious about seed should make a plantation on a raised bank of poor sandy soil, in a bleak situation, to increase the seed-bearing tendencies of the plants. In fat soils they rarely produce seed, and are likely to be short-lived.

VIOLA PEDATA.

CHAPTER VIII.

TENDER BORDER FLOWERS.

The plants classed in this section are such as require to be raised every year from seeds under glass, with, in most cases, the aid of artificial heat; or to be preserved with particular care during the winter, and have the aid of heat to start them into life in spring. They are distinguished from hardy perennials and hardy annuals by the fact that they are so far tender in constitution that it is only during the summer months they can endure exposure to the common atmosphere. Fortunately for the cultivator, they readily adapt themselves to a variety of circumstances, provided only they are warm enough, and for the most part they are rapid-growing plants; so that, very soon after being planted out they attain their full stature and flower freely. To speak of tender border flowers in a comprehensive manner, we might say that the instructions offered on the cultivation of bedders apply to them with but trifling exceptions, which the amateur will soon discover for himself. But our duty is to be particular and precise, however brief; and therefore we shall again attempt, as in previous chapters, to provide very short but thoroughly practical codes of management for the several subjects that claim attention here.

Though much may be done by means of cold frames, and by economizing spare corners in a greenhouse or early vinery in the growth of tender border flowers, the amateur who would do things well must encounter the few difficulties that attend the construction and management of a

Hotbed.—To heap up a quantity of stable manure is one thing; to make a serviceable and lasting hotbed is another. The method of procedure must in some degree depend on the nature of the materials at command for the purpose. Stable manure that has been slowly accumulating in a heap, and the greater part of which is in a half powdery condition through fermenting long and furiously, will make a first-rate hotbed, with an extremely small amount of trouble. It is only necessary to make up a bed three feet deep, and large enough to extend two feet beyond the frame every way, and there is a hotbed at once mild and lasting. In making the bed, see that the manure is moderately moist throughout; if dry and flaky, and, perhaps, blue with mildew, throw water over it as the work proceeds—not in such a way as to saturate one portion and leave another dry, but to make it moderately moist throughout. Put on the frame, and then cover the manure inside the frame with six inches of good soil, consisting of turfy loam from a heap of top-spit turf that has been laying by for a year, with a good proportion added of old hotbed manure rotted to powder, and sharp sand, to render the mass porous and light. Road-scrapings from gravel roads are to be preferred to pit sand; or the siftings of the sweepings of gravel walks answer well for the amelioration of a good loam in preparing a seed-bed. If leaf-mould is obtainable, it may be employed to great advantage, mixed with turfy loam, to cover the manure as a bed for the plants.

We must suppose, however, that fresh manure only is obtainable; and in this case it must undergo a systematic preparation, for if heaped up in a crude state, it will ferment so fiercely as to burn up seeds and plants, and ruin any and every enterprise. Let the manure be well shaken out, and laid up in a heap as lightly as possible, and, if dry, sprinkle water on it as the work proceeds. In the course of about four days shift the whole mass to another spot, breaking all the lumps with the fork, and lay it up again. If it happens to be short and pasty, as it will be if there is any considerable proportion of it drawn from the cow-byre or pigsty, mix with it straw, fern, old turf, or other dry vegetable litter. When it ceases to ferment furiously, and has acquired a steady heat, make up the bed as directed above, in the use of manure already much fermented, except that in this case the bed should be full five feet deep. In any and every case a mere handful of stuff is of no use. To be sure, an experienced hand can do much with poor materials, and a one-horse load of good stable manure will suffice, under good direction, for a hotbed that will stock a garden with dahlias, asters, balsams, and many other things. But, as a rule, any less quantity than four horseloads is useless; and so we advise the beginner

to begin by making a good bed, and to wait until experience
has taught how shifts may be made, and severe economy
practised. Moreover, a well-made hotbed will abundantly
pay for its cost, for after it has supplied seedling plants for
the flower garden, it will be in a good condition for growing
marrows or mushrooms, with a proper improvement of the
top soil by the addition of good loam and manure, according
to the requirements of the case. The larger the bulk, the
longer will the heat continue, and the more steady will it be.
When the frame is put on, it is probable that the heat will
rise to too high a pitch, in which case the frame must be
tilted to allow the steam to escape. The beginner must bear
in mind, that if the whole affair is as light as possible, the
heat will be more moderate than if it is pressed or beaten
down ; therefore, in the employment of rank manure, though
two or three times turned, care should be taken not to tread
on the beds more than is absolutely necessary. On the other
hand, if old manure is employed, and the heat does not rise
as desired, tread down the manure pretty firm before putting
on the soil, and there will soon be a nice heat generated that
will last long enough, with a careful husbanding of the
warmth derived from the sun, by shutting up early, and
giving no air at all on bleak, dull days. It is always better
to sow seeds and to strike cuttings on a bed in a frame
over a mass of fermenting material; nevertheless, pots and
seed-pans may be employed instead, or both systems may
be pursued simultaneously. Our practice, for many years
past, has been to make up a bed with about twenty or
thirty loads of well-rotted manure, and put on the frames,
and set them to work at once, regulating the heat by judicious
ventilation. The bed is kept at work throughout the summer, for various purposes, and in the winter is cleared away,
and the manure stored in the potting-shed, to be ready for
use in preparing composts, and to make the ground ready for
a new bed in the spring. When propagating on hotbeds
is commenced early, it is necessary to have ready in good
time a second set of beds, on which to prick out the plants
raised in the first, because tender subjects must be *kept growing*
until they can be safely planted out. The amateur who grows
but a few choice plants, and has but few conveniences for the
pursuit, will do well to defer to the *latest moment* possible the
commencement of hotbed work, because then the sun will be

helping every day ; and very soon after the plants have grown
to a size large enough to be handled, they may be planted
out, and nature will take kindly charge of them.

" So manifold, all pleasing in their kind,
All healthful are th' employs of rural life,
Reiterated as the wheel of time
Runs round ; still ending, and beginning still."

ASTER.—The aster is commonly and properly designated a
" half-hardy annual." We class it here with tender border
flowers, because its requirements assimilate closely with a few
perennial plants of similar constitution. To grow it well, it
must be grown quickly, and never suffer a check from the
first. If carelessly treated, it becomes the prey of green-fly
and red-spider ; or, if these haply abstain from assailing it,
starvation marks it for her own, and a yellow leafage and a
shrunken flower tell surely of the hardships it has endured,
and it cannot prove the joy of the garden, as with good
treatment it will surely be when its season of flowering
arrives. Beginners are apt to sow the seed too soon, and so
involve themselves in trouble ; for the instant that the plants
are large enough, they should go to the open ground, and
there have encouragement to grow freely, exposed to all the
winds of heaven, with water only to help them through times
of drought. The first step towards a good display of asters
is to obtain the best seed possible, and as there is plenty of
bad seed in the market, the purchase should be made from a
house of known respectability. Home-saved seed is worthless, so do not trust to it. From the end of April to the middle
of May is the proper time to sow the seed, and it is well to
promote germination on a mild hotbed, or by placing the
seed-pans in a greenhouse. They may, however, be very well
started in a cold frame, if kept closely shut up and carefully
managed as to air-giving after the young plants appear. It
is good practice to plant out the stock as soon as the plants
are large enough to handle—say when they are an inch high—
on a nearly-exhausted hotbed ; the object being to promote
a quick growth. But a bed in a cold frame will serve the
purpose, and they must have as much air as can be given them
with due consideration of their tender nature and the state of
the weather ; when they are three inches high they should be
planted out where they are to flower. If required simply

12

to make a gay border or bed, any good garden soil will suit them, and they should stand six inches apart. But if fine flowers are required they must be planted out on soil well dug and liberally manured, quite a foot apart each way, and when planted a thin coating or mulch of rotten manure should be spread over the ground amongst the plants. If well managed from the first, they will not need the support of stakes; but the cultivator must determine this point, and if it is needful to assist them with the support, they should be staked neatly some time before the flower-buds begin to swell. Having grown hundreds of thousands of all the sorts known, we have found it not only a saving of time but the better for the plants that they should not be staked at all; but if they are drawn in the early stages of growth, or are peculiarly exposed to strong winds, they must be assisted. In dry weather they should be well watered, and if flowers of high quality are desired, the flower-buds should be thinned to three or four to each plant as soon as they are visible. Slugs are great enemies of asters, and where these pests abound it is a good plan to plant lettuces in the beds at the same time as the asters, both to decoy the slugs from the asters and also enable the cultivator to crush the enemy; for they will congregate about the lettuces, and may thus be caught night and morning, and it will be well to hunt for them after dark by means of a lantern. Red-spider and aphis are terrible destroyers of asters. To prevent their coming keep the plants growing freely, for it is the starving plant they search for and love; the strong plant is not to their taste at all. Occasional dusting with dry powdered lime or sulphurized tobacco-dust will be found of great service when asters are assailed by any of these destroyers, but the golden rule from first to last is to insure a vigorous growth and a state of robust health, and then even the unfastidious slug will scarcely care to touch them, for he, like the rest of the marauders, has a special love for a soft sickly plant. A pleasing display of asters may be obtained by sowing the seed, on the spot where the plants are intended to flower, about the 10th of May, and thinning the plants to six inches apart   This simple system produces only late flowers of inferior quality. To insure fine asters the plants must be cultivated.

The best varieties of asters are the *Pæony-flowered*, *Chrysanthemum*, and *Quilled*; of each of which there are several colours. In all cases the dwarfest sorts make the most complete and beautiful masses, and the tall ones are the most useful for cut-flowers.

BALSAM.—This noble flower requires the same general treatment as the aster, and is a few degrees more tender in constitution. As a bedding plant for any odd position it answers well, and may be dealt with in a rough-and-ready fashion. Prepare the ground by deep digging and liberal manuring, and sow the seed thinly about the 20th of May. Thin the plants to full two feet apart, help them with weak manure-water, and they will soon cover the bed and flower splendidly, unless the season happens to be unusually cold, in which case the bed of balsams will be a down-right failure. To grow fine balsams, sow on a hotbed the last week in March, and again in the last week of April. Warmth and moisture are most important aids in the culti-vation of the balsam, for it should grow fast from the first, and never suffer a check. Prick out the plants from the seed-pans when their seed-leaves are fully developed, putting them in light rich soil on a hotbed, the heat of which is never lower than 55° to 65° at night. Plant them so deep that their leaves almost touch the soil; sprinkle them frequently with water of the same temperature as the air of the frame, and ventilate carefully to promote a sturdy habit without checking the growth. Frequently lift and plant out or pot them in rich soil, so as to afford the roots more and more room, and keep them growing fast in a frame over a nearly spent hotbed until they become great bushes, when they may be allowed to flower. They may be grown to almost any size if the flower-buds are constantly picked off until the plants are as large as required. A few fine balsams in pots are of the greatest value to embel-lish the greenhouse and sitting-room in the height of summer, and in the process of producing them, the least promising plants will be found useful for planting out in beds and borders, but they must not be put out until Midsummer-day unless the season and the situation are both peculiarly favourable. The best varieties are the *Rose-flowered* and *Camellia-flowered*, but worthless seed is commonly sold under these names, and the only way to insure seed worth growing is to go to a house *known* to be trustworthy, and pay a good price for it. The more perfect a balsam is in form and colour the less productive will it be of seed, but trashy balsams will produce abundance,

and hence those who vend inferior articles experience no
trouble whatever in making up low-priced packets of balsam
seed.

DAHLIA.—As a border flower the dahlia is certainly worth
the little care it requires, but it is not a first-class border
plant, though, if regarded from the florist's point of view, it
is one of the grandest flowers of the garden, and in rank must
be placed second only to the rose. When required to form the
background of a plantation, intermixed with hollyhocks,
aconites, and other tall-growing plants, it is only necessary to
put out in the common soil roots that have been stored in sand
the first week in May, or wait until the first week in June, and
then put out young plants that have been carefully hardened
in a frame. The bouquet dahlias are especially valuable for
the mixed border, because their comparatively small flowers are
produced in great profusion, and they are more useful as cut-
flowers than those of the exhibition class. When the frost has
cut down the plants, the roots should be taken up with a few
inches of the stem attached as a handle, and be stored away in
sand in a loft or some other cool dry place. To grow the
dahlia with a view to the production of fine flowers something
more must be done than this rude code requires. The roots
are started into growth on a hotbed or over a tank in a warm
greenhouse in March, and if a large stock is required the
shoots are taken off and struck in heat as fast as they can be
obtained. But if only a few good plants are wanted, the first
lot of shoots are broken off and thrown away, and the second
lot are struck; these making better plants than the first.
They must be kept growing freely in the fashion of bedding
plants, and be hardened off in like manner for planting out.
The plantation should be made on a piece of ground that was
prepared for the purpose in the previous November, by trench-
ing and manuring. It should lie open to the south, but have
the shelter of trees from the north. A shady or confined spot
will not do. It is a common mistake to plant early in order to
obtain extra growth and early flowers. Early planting is a
needless exposure of the plants to a thousand baneful in-
fluences. The first week in June is the proper time to plant,
but some time in May, and the sooner the better, the plot
should be planted with lettuces, and these should be constantly
hunted for vermin. The proper way to plant is to open
holes five feet apart, and dig in some good rotten manure to a

SHOW DAHLIA.

(Crimson self.)

depth of two feet. Then plant carefully, filling in round the plant with fine earth, and drive down a strong stake behind the plant about four inches distant from it. Finally drive in two shorter and lighter stakes in front of the plant, about eighteen inches distant from the stake in the rear, to form a triangle. As soon as the plant is tall enough tie it to the main stake, and pass the matting on either side of the plant to the stakes in front to form a sort of cage. The farther tying will be a very simple matter. In dry weather copious supplies of water must be given, and by the middle of July the roots should be mulched with good half-rotten manure. The earwig will now begin to make its mark on the plants and must be trapped. For this purpose there is nothing to equal small flower-pots, each containing a bit of dry moss or hay, and mounted on the top of the principal stake *above the plant.*

Dahlias vary very much in growth, and therefore need variations of treatment. Those that make over-much growth must be thinned so as to allow free access of light and air to the principal branches. Those that present a great number of flower-buds must be disbudded in order that the flowers may be of good quality. In removing shoots pinch them out when very young; and if uncertain about the extent of thinning required, take care to err on the side of leaving the plants rather too crowded, than to reduce them in a degree detrimental to their vigour. The shading, dressing, and exhibiting of the flowers are subjects that do not properly claim attention here, but we subjoin a list of first-class varieties that are likely to be considered good until ·1880, and perhaps a year or two beyond.

SHOW DAHLIAS.—BEST FIFTY.

Light : *Julia Wyatt, Mrs. Brunton, Hon. Mrs. Wellesley, Unique, Queen of Beauties, Heroine, Dawn, Mrs. Dodds, Miss Henshaw, Peri, Anna Keynes, Alexandra, Princess, Harriet Tetterell, Flag of Truce, Adonis, Heby, Lady Derby, Caroline Tetterell.*

Yellow and Orange : *King of Primroses, James Hunter, Samuel Naylor, Chairman, Hugh Miller, Mr. Boshell, Charles Turner, Fanny Purchase, Leah, Lady M. Herbert, Vice-President, Royalty, Toison d'Or.*

Crimson and Red : *Mr. Dix, Triomphe de Pecq, British Triumph, Bob Ridley, Sir Greville Smythe, Aristides.*

Purple and Maroon: *Indian Chief, Andrew Dodd, Lord Derby, George Wheeler, James Backhouse, Paradise Williams, High Sheriff.*

Lilac and Rose: *Memorial, Juno, Lilac Queen, Criterion, Mrs. Boston.*

FANCY DAHLIAS.—BEST TWENTY-FOUR.

Striped and Spotted: *Lady Dunmore, Madame Nilsson, Purple Flake, Octoroon, Regularity, Sam Bartlett, Ebor, John Salter, Artemus Ward, Butterfly, Grand Sultan, Leopardess, Viceroy.*

Dark Tipped: *Polly Perkins, Lady Paxton, Mrs. Crisp, Nora Creina, Pauline, Pluto, Queen Mab, Prospero, Fanny Sturt, Gem, Viceroy.*

BEDDING DAHLIAS.—BEST EIGHTEEN.

Light: *Queen of Whites, Alba floribunda nana.*

Yellow: *Duke of Newcastle, Golden Bedder, Golden Ball, Leah.*

Scarlet: *Beauté de Massifs, Scarlet Tom Thumb, Rising Sun.*

Rose and Lilac: *La Belle, Rose Gem, Scarlet Gem, Blonde.*

Crimson and Purple: *Tom Thumb, Crimson Gem, Royal Purple, Zelinda, Floribunda.*

LOBELIA.—The magnificent plants known in gardens as "herbaceous lobelias," descendants of *L. cardinalis, L. fulgens,* and others, have never enjoyed the favour to which they are entitled, though at the present time they are comparatively unknown, as compared with the partial recognition of their merits a quarter of a century ago. The garden varieties are the perfection of border plants, and a few amongst them having distinctive purple, bronze, or claret-tinted leafage, as well as brilliant flowers of divers hues, may be employed as bedders with eminent advantage. In the cultivation of these fine plants some little skill and care are necessary, and there are two ways of managing, which may be termed respectively the gardener's and the cottager's methods. As the cottager's is the most simple, let us begin with that. Some time early in the month of May a few plants are purchased and planted on a deep well-manured border in the full sun. If they are to make a mass, they may be a foot apart, but much better to put them in clumps of three each in the midst of lupins, delphiniums, lychnis, and other such bonny "old-fashioned" flowers, for the many tints of green there are in such borders, and perhaps the fine deep shadows of shrubs and trees in the rear, help to bring out the colours of these noble lobelias. To bring them to perfection by this treatment, all that is further needed is abundance of water. Give each clump half a gallon every evening (except during rainy weather), from the time of planting till the first flowers open; and then discontinue watering, as the season will be advanced, and showers will probably suffice for their wants. As soon as the bloom is over they should be taken up, and be potted in light soil, and be kept out of doors till the nights are frosty, when they must be housed in a pit or greenhouse, or a window somewhere safe from frost, but in full light; and all winter they must have air as often as possible, and a little water to prevent them getting dry. Mind they never go dry, winter or summer—it is a *golden rule*. As for propagating, you can divide in May when planting out, or you can sow seed in May and June, or you may strike cuttings in autumn or spring if you can give them a little heat, or even without heat if you know how to strike cuttings at all.

Now for the gardener's system, which requires glass and grand notions. Gardeners are oftentimes puzzled to know how to vary the July and August show when all the "good things" are over. Let them try lobelias of the cardinalis section, and cry out again when they have mastered all the points in the cultivation of these glorious subjects. Supposing the plants to be purchased in March, they ought at once to have a shift to 32-sized pots, the compost to be silky loam, leaf-mould, turfy peat, and rotten manure, equal parts. In these pots allow them to flower in the greenhouse, giving abundance of light and water until the first blooms open, and then gradually diminishing the supply. They will be useful in the conservatory, and will show their qualities sufficiently to prove that if well grown another season a sensation might be made with them. While they are in bloom mark the best for specimen growing, and at the end of October begin the routine. Take off the suckers as soon as they can, be removed with something like a heel to them. Pot them in five-inch pots singly, and plunge in gentle bottom-heat. Use the same compost as above recommended, with one part of silver-sand added; in after shifts return to the original compost, omitting the silver-sand. Early in January shift into six-inch pots, and put them in an early vinery, or wherever the heat averages 50° by night and 60° by day. In a month's time shift again to eight-inch pots, and give them a rise of 5° to 10' in temperature. About the middle of April shift again to ten or twelve-inch pots, put them in a cool house slightly shaded, and give abundance of water. When the spikes appear, put in light stakes five feet high, and tie the spikes in carefully from the first, to prevent them getting bent or

twisted, and for the rest— wait and see. Of course the two systems can be combined, and all plants not wanted for indoor display can be put out in rich mellow borders in the month of May to take their chance.

In selecting for ordinary purposes, the garden varieties are the best. But such distinct species as *L. cardinalis*, 3 feet, scarlet flowers; *L. syphilitica*, 2 feet, light blue; and *L. nicotianæfolia*, 6 to 12 feet, flowers pale lilac, are invaluable, the last being admirably adapted to stand alone during the summer in a sheltered nook, where its noble outlines would be seen to advantage. The following are the names of a dozen varieties of different colours, and the finest quality, for decorative purposes :—*St. Clair*, crim-

LOBELIA NICOTIANÆFOLIA.

son, a fine bedder; *Carminata*, carmine; *King of Blues*, blue; *Alba grandiflora*, white, with blue veins; *Ceres*, rose; *Sappho*, reddish-purple; *Distinction*, cerise-red; *Nonsuch*, violet and vermilion; *Ruby*, ruby; *Excellent*, magenta; *Peach Blossom*, peach and vermilion; *Victoria*, rich scarlet, a fine bedder.

MARVEL OF PERU.—"Is it worth growing?" Oh, ingrate world, to ask such a question! Look into the tiny front court

of the cottager in the cool of the day, and be struck dumb
with astonishment at the scintillating beauty of the great
round dense bushes clothed with bright light green shining
leaves, and pink, white, scarlet, purple, or rosy flowers, that
truly glitter as if blessed with a better sort of daylight than
the grand plants that swelter in the sun in my lord's garden.
That brilliant buxom thing is the Marvel of Peru, a marvel
to me and you; the botanists call it *Mirabilis jalapa;* the
specific name suggesting an unpleasant experience of the
youthful palate. The plant has small carrot-shaped roots,
which are kept in sand during winter, and are planted out in
April or May. If the soil is deep and rich, there never need
be a drop of water given, and the growth is so orderly and
self-dependent that sticks and ties are never required. To
raise them from seed, sow on a hotbed in spring, and plant
out the seedlings in the early part of June.

STOCK.—Six several chapters might be written on the cultivation of stocks, showing how to manage them as pot plants,
as bedding plants, for early bloom, for late bloom, and for
raising seed and new varieties. Our duty is to avoid those
matters that pertain to the nursery and the market-garden,
and provide directions for the employment of the stock as a bedding and border plant simply, and a very few words will suffice.
In the first place we condemn *in toto* all troublesome and
complicated methods of procedure, because they necessitate a
wasteful expenditure of time, and actually tend to the production of flowers that are bad in proportion to the time wasted
upon them. In the next place, we denounce as sheer foolish-
ness all the rules proposed for distinguishing double from
single stocks when the plants are in a small state. To a certain extent it is in the cultivator's power to make them all
double, and our simple code of culture will indicate the proper
order of procedure. Home-grown seed is rarely of any use;
indeed, the production of good seed is an art demanding more
skill and patience than any average amateur can devote to it.
Secure the best seed possible from a first-class house, and sow
it on any day between the 10th and 30th of March. The seed-
pans or pots should be filled with light rich soil, consisting of
about equal parts of leaf-mould, rotten manure, and sandy
loam. The proper place for the seed-pans is a cold frame, and
it will be well to lay slates, tiles, or sheets of glass over them,
to assist germination and render watering unnecessary. If

the soil becomes rather dry, however, it must be carefully moistened with the syringe, or by dipping the pans into a vessel containing a sufficient depth of soft tepid water. Instantly upon the plants appearing remove the covers and let them have light and air, the ventilation being regulated by the weather so as to render the plants as robust as possible without causing a chill. They are not to be pricked out to strengthen, nor are they to be kept in the seed-pans to starve. As soon as they are large enough to handle, which they will be by the middle of April, they must be planted out and encouraged to grow freely from the very first. Any soil that will grow cabbages will produce first-rate stocks, but it should be deeply dug and liberally manured long in advance of the day of planting: better indeed, if prepared expressly for the purpose some time in the winter and left rough to the last moment. It is, however, not absolutely necessary to prepare the ground until the last moment, but it must be well done, and the manure, in liberal quantity, thoroughly well broken up with the staple soil. When the digging is finished, spread over the bed two inches of manure rotted to powder, and prick it in with a small fork. Then draw drills fifteen inches apart and two inches deep, and in these drills insert the plants three inches apart. An experienced workman would lift the plants out of the pans by the aid of a bit of stick, and lay them in bunches towards the left hand, and presenting one between the finger and thumb, make a hole with the right hand, thrust the plant into it with the left, and close with the right, at a rate of speed which would astonish a novice looking on.

If frost should follow, the little plants must have some kind of protection, and there is no more speedy and effectual method of providing it than to cut a lot of short branches of spruce, or any evergreen that can be spared, and stick them all over the bed. A net spread over, and kept from touching the plants by means of a few stout stakes, will answer well. Water must be supplied in plenty during dry weather, and as fast as the growing plants touch each other thin them out, always removing the weakest and those that show flowers first. When there begins to be a show of colour all over the bed, make a final thinning, taking out all that present single flowers first, and then the forwardest of the double ones, until the plants are far enough apart to promote full development, and yet not too far for a rich effect. When the thinning is completed give the bed a good soaking with weak liquid manure, then carefully touch it over with a small hoe or rake to make a neat finish, and the routine of your cultivation is completed. You may now, send out cards of invitation to friends, for you will have a bed of stocks that will be worthy of admiration, and far too good for your own enjoyment solely.

Stocks may be sown in September and wintered in frames for an early bloom. They may be sown again in heat in January or February, but by no other course of cultivation than we have here described, is it possible to obtain them in perfection. The best sorts are *Ten Weeks* for summer display ; *Intermediate* for autumn. The *German Dwarf Bouquet* and *German Large-flowered Pyramidal* are useful. A few good sorts are to be preferred to anything like a collection ; indeed, collections are only adapted for experimental gardens, the directors of which expect always to bestow their time on many things of quite secondary value.

ZINNIA. — The habit and requirements of this plant so closely correspond with those of the aster, that they may be grown side by side from first to last, and there will be no shadow of difference in their behaviour. The double Zinnias are magnificent when well grown, but the single varieties cannot be dispensed with. A set of beds on an open sunny lawn devoted severally to asters, balsams, stocks, and zinnias, or one great bed containing a mixture of them, would afford the frequenters of the garden a rare and delightful entertainment, for few people grow these charming flowers, owing to the prevalence of a false faith in geraniums and verbenas as the only plants that can be persuaded to flower out-of-doors in any garden in Great Britain.

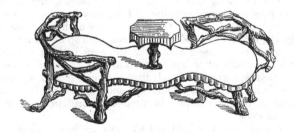

CHAPTER IX.

BEGINNERS take to annuals with peculiar fondness, but as they acquire experience their love for the friends of their youth declines, and they soon become indifferent to annuals, even to the extent of abusing them as weedy short-lived things. Many of the hardy annuals *are* weedy and short-lived ; some are exquisitely neat and gay, and also short-lived ; a few are equal in beauty to any perennials known whether hardy or tender, and last as long in the gayest trim as any one can desire who can find in the changeableness of plants a greater source of pleasure than could possibly be found in unchangeable beauty, were they so unfortunate as to be like cast-iron or the cold hard work of the sculptor. If we are not to despise the day of small things, we must make room in the garden for a few hardy annuals, and it will soon be found that they have some peculiar claims to regard, which we will endeavour here to state in very few words. To begin with, they are cheap, and any one can grow them : those two reasons, perhaps, prevail with beginners. They are exceedingly gay, and the best of them last long enough, considering that by proper management a long succession of flowers may be obtained. They may be wholly grown from first to last without the aid of glass or flower-pots, or composts, or sticks, or shades, or even a drop of water, and will yet make a liberal return for the very little care which their simple cultivation requires. There is no other class of plants that can give an equal display of colour and an equal range of characters and colours, gay and various, for the small amount of labour required to produce a brilliant border of hardy annuals.

We shall first speak of the simplest mode of cultivating these plants. We will suppose a sunny border, and it may be a few beds in a sunny situation, and the month of February has come, the spade must be employed to dig deeply and break up the ground well. If a good dressing of half-rotten manure can be dug in, the result will be a more brilliant and lasting display than can be insured by withholding manure, but even that is not absolutely necessary. The finishing touch should be given to the border by thoroughly breaking up the surface soil to produce a fine seed-bed. The best way to sow the seeds is in patches of one to three feet across, and the same distance apart as the size of each patch. It must depend upon the size of the border as to the distances and sizes of the clumps, but a few large clumps will be better than many small ones, even on a border of most limited dimensions. Now we come to the sowing of the seed, and have to say that the seeds of hardy annuals are always sown too thick, and there are always too many plants left in the several clumps. Good reader, kindly bear in mind for your own joy, that one plant of the common Virginian stock allowed to attain complete development will cover more than a square foot of surface, and produce flowers as large as a florin, and last two months in bloom; while if twenty plants occupy the same space they will be spindling weedy things with flowers the size of threepenny bits, and all over in three weeks at the utmost. The one grand secret in securing a fine bloom of hardy annuals is to sow early, and thin severely, and to proportion the thinning to the growth of each sort, so that every separate plant in a clump shall have room to spread and be encouraged to make much growth before it begins to flower. The time for sowing seed is February and March, and the surface soil should be fine and dry when the work is done. The seed should be thinly scattered in the circles allotted to the several sorts, and be covered with finely-sifted earth, about one inch deep generally speaking, but the larger seeds may be dropped into holes made with the finger or a stick, and the larger they are the deeper they should go; those of lupins, for instance, may be two or three inches deep, the little seeds of Virginian stock, on the other hand, should be but just covered. It may be that bad weather prevents early sowing, in which case the month of April remains for a last opportunity, and a very good display may be obtained in June and July by sowing even so late as the end of April or the first week in May. But as to the advantage of early sowing there cannot be a question, for the longer the period in which the plants grow and spread before they flower the finer at last

will their flowers be; for late-sown seeds are hurried into
flower by the heat of the sun before a fair sized plant has been
formed, and the bloom is rich or poor in proportion to the
strength or weakness of the plant that produces it.   There need
be little fear of cold weather destroying early-sown seeds, for
we find that seeds of all kinds, including many of the most
tender plants, remain dormant and unhurt all the winter, and
indeed until the season has advanced sufficiently for their safe
emergence, and then they grow with their proper vigour and
the resultant plants not seldom surpass those that have been
nursed under glass with tender care.   There is, however, this
risk in early sowing, that warm weather may promote germination and cold weather may follow and kill the plants.   After
making a fair balance, we conclude to advise the practice of
sowing early, the advantage on the average being so great as
to render the necessary risk a matter of comparatively small
consequence.

From the first appearance of the young plants, thinning
and weeding must be regularly performed.   The ground may
be occasionally chopped over with the hoe to keep the surface
open to sun and shower, but excessively careful raking, intended to make the surface as fine as snuff, is to be avoided
as a waste of labour for a bad result, and watering is to be
avoided too, unless the soil is poor, and the weather unusually
hot and dry, in which case a plentiful supply will help the
plants greatly.

Some annuals, as the sweet pea for example, may be supplied with light stakes for support, and others, as the annual
chrysanthemums and scarlet flax, may be neatly trained to
light stakes.   But all may be grown without artificial support,
and wherever it can be dispensed with, we should save the
time, and obtain a more pleasing display, than by the most
careful staking.   This dictum must be taken *cum grano*.   We
are no advocates for slovenly gardening, but for promoting
the highest development of everything taken in hand, and
allowing each plant to express its character with as little
interference as possible.   Even sweet peas, rapid climbers as
they are, may be allowed to trail over a sunny knoll without
a stick to help them, and, if they have but room enough, will
make a beautiful rustic sheeting of healthy leafage and brilliant flowers.

In the display of annuals there is the same room for the

exercise of taste as in the display of any other kinds of flowers. The amateur may desire to have great variety, or may prefer a few of the very finest sorts, and repeat them again and again to produce a rich effect. In a garden frequented by a few interested observers, the first plan might afford the most continuous and changing pleasure : in a garden frequented by many, where plenty of colour is generally a matter of first importance, the second plan would be the best.

The simple system thus far sketched out is the best for general purposes, but for special purposes other plans may be adopted. Thus we may sow all kinds of hardy annuals in August, and a large proportion of them will germinate at once, and make strong plants before winter, and bloom earlier and stronger than those sown in spring. In this case a somewhat poor and dry soil should be chosen, but really it matters very little if the soil is cold and damp, for do we not see on the worst of soils, and on the best alike, self-sown wallflowers, mignonette, sweet peas, candytufts, and many other things that have managed their own affairs in their own way, the seeds having been shed in July, germinated in August, become little green bushes by Christmas, and bonny flowering plants in the month of May. In one part of our garden mignonette is an established weed, and we have every year to thin out the self-sown plants to six inches apart, or they would crowd each other to death. The common wallflower haunts us in the same manner, and we have to destroy hundreds every year where this takes place. The soil is damp and cold, the aspect north-east, and the bleakest anywhere within half-a-dozen miles of the Bank of England, and the border is heavily shaded by large trees. To sow annuals in autumn cannot, therefore, be so mad a procedure as some people profess to regard it ; but perhaps they do so regard it because when they have tried the experiment they have been *too late*, and the miserable rains of October have caught their poor plants just coming up, and have killed them clean off, as a silent and bitter reproach for pretending to follow the book when proceeding dead against its advices. We name the month of August for sowing annuals to stand the winter, but in the north July will be none too soon, whereas everywhere June would be too soon, because June sown annuals will, if they can (this is, you know, weather permitting) flower nicely in

September and October. Thus you see we glide into another expansion, and our only way of disposing of a great matter in few words is to say that June is not a good time to sow annuals of any sort. But a good sowing may be advantageously made at the end of May for late summer flowering, and they must be kept liberally watered to prolong the season of growth before flowering begins.

Next we expand the scheme into frame and pot culture. In great and grand gardens the cheapest and commonest annuals are grown in pots for the embellishment of the conservatory, and most beautiful are the tufts of nemophila, schizanthus, and leptosiphon so produced. We are quite among the fine arts now, and must beware of expanding this chapter beyond reasonable limits. But it may be remarked that as variety is charming everywhere, any greenhouse or conservatory may be prettily embellished with annuals in pots in the early months of the year, and especially at the time when somewhat of a clearing out is made, and the house is rather bare of embellishment, for camellias and acacias will be past at the time the annuals flower, and they will contribute in a most agreeable manner to the providing of a gay garden under glass at a time when flowers are looked for, and there are as yet but few in the open ground.

The advices offered on the saving of seeds of perennial plants apply strictly to the saving of seeds of annuals. The best general advice we can offer on the subject, however, is that seeds should not be saved, but should be sedulously removed as fast as they are produced, both to preserve the order and brightness of the garden, and to prolong the display of flowers. By carefully picking off all seed-pods the instant the flowers fall from them, the plants will be encouraged to continue in flower to the very end of the season, or, if they do not hold out so long, it is very certain that twice as many flowers may be obtained from any of them if the development of seeds is prevented by constant suppression of them, than in the opposite case of their being allowed to swell and ripen naturally.

The amateur must always bear in mind that the multiplication of annuals need not depend on seeds alone. Every one of them may be multiplied by cuttings in precisely the same manner as we have advised in respect of sweet-williams and wallflowers. It is a question, of course, if it is worth while, in any particular case, to resort to this method. Generally speaking, seeds are to be preferred.   But there will occasionally occur in a bed of annuals a plant, or many plants, presenting distinct and desirable characters—it may be double flowers or variegated leaves, or flowers of a different colour

GLOXINIA-FLOWERED FOXGLOVE.

to the ordinary type—and the question will arise, Shall we ever attain the like if we trust to seed?   In such a case, if the thing is worth keeping, the blind shoots at the base should be removed and struck, and a stock secured for future

13

use. This should be done on the instant of the discovery of
the value of the plant; there must be no waiting until the
flowering is past, for then it may be forgotten, or the plant
may die. The practised cultivator, who has a taste for keeping a "good thing," would indeed at once cut off the head of
the plant, and sacrifice the flowers, in order to obtain a free
growth of young shoots, making sure, too, of a few to begin
with, and having in view to cut and come again.

But what about Biennials? These may be disposed of in
a general way, by saying that they are in all respects the
same as annuals, but usually do not flower until they have

SUMMER-HOUSE IN CHINESE STYLE.

passed through one winter; and having flowered, they generally die; and therefore, like annuals, have to be renewed
from year to year. It is impossible to classify garden plants
strictly as annuals, biennials, and perennials; for some so-
called annuals will live through the winter and flower again,
some so-called biennials will flower the same season that they
are sown, and very many will do so if they are sown early
on a gentle hotbed, and are coaxed along in frames, and
are planted out when they have attained to a good size. And
again, some so-called biennials last many years, and become

veritable perennials. The selection that follows includes such species and varieties as for their beauty or perfume are desirable in every garden where annuals are grown. The list may be extended immensely if "all the good things" are included, but for general purposes a short list is better than a long one, and almost every plant named will suggest to those who take an interest in them, others of the same genus or species, that are equally worthy of cultivation, but which are omitted simply because we must not only begin somewhere, but also make an end somewhere.

A SELECTION OF HARDY ANNUAL AND BIENNIAL FLOWERS.

FRONT ROW (averaging six inches high).—*Asperula azurea*, lilac and blue. *Campanula pentagonia,* purple. *Collinsia Bartsiæfolia alba*, white. *Eschscholtzia tenuifolia*, yellow. *Gilia tricolor*, white and purple. *Godetia reptans*, white and purple. *Kaulfussia amelloides*, beautiful blue. *Leptosiphon androsaceus*, lilac, and *L. densiflorus*, purple; *L. roseus*, very beautiful rose. *Reseda odorata*, sweet-scented mignonette. *Malcomia*

LEPTOSIPHON ROSEUS.

maritima, Virginian stock, white and rose. *Nemophila insignis*, beautiful pale blue, and *N. insignis alba*, white. *Oxalis rosea*, exquisitely beautiful, bright rose, a tender plant, flowering late when sown on the border. *Portulacca*, various, exceedingly brilliant when grown on a dry, sunny, sandy knoll. *Sanvitalia procumbens*, yellow. *Saponaria calabrica*, pink; *S. calabrica alba*, white—two of the gems of the annual border. *Silene pendula*, rose, a delightful plant.

SECOND ROW (averaging 12 to 15 inches). — *Brachycome iberidifolia*, a neat bedding plant, blue. *Calliopsis Englemanni*, golden. *Calandrinia Burridgi*, rose: this represents a splendid family. *Campanula Lorei*, blue; extremely pretty. *Chrysanthemum carinatum*, yellow and brown; *C. flavum*, gold; *C. venustum*, purple and yellow: a fine group. *Clarkia pulchella*, rose, and *C. integripetala*, rosy crimson, represent a fine family, which should be grown in plenty. *Collinsia bicolor*, purple and white; *C. multicolor*, crimson and white; lovely plants of free growth and abundant bloom. *Delphinium ajacis*, Larkspur, white, pink, blue, and purple; the New Dwarf Rocket and Candelabra Larkspurs are the best for the second row. *Delphinium cardiopetalon*, blue. *Dianthus chinensis* in great variety, all of them splendid plants for masses, making a good show when sown on the open border. *Eschscholtzia Californica*, yellow, and *E. crocea*, orange, showy and neat, apt to become weeds like mignonette and cornflower. *Eucharidium grandiflorum*, red. *Eutoca viscida*, blue. *Gilia achillæfolia alba*, white; *G. capitata*, blue. *Godetia Lindleyana*, purple. *Gypsophila elegans*, white, fine for bouquets. *Helichrysum elegans*, fine yellow everlasting. *Iberis umbella*, the Candytuft, in variety; the white, crimson, and purple are splendid things when well grown, but they are usually ruined for want of thinning. *Ipomœa tricolor*, dwarf convolvulus, white, blue, lavender, etc.; these make fine tufts, and answer well to cover the ground among standard roses—the flowers always look to the south. *Limnanthes Douglasi*, white and yellow. *Linum grandiflorum*, crimson Flax, a splendid plant. *Lupinus affinis*, blue and white; *L. luteus*, yellow and sweet-scented; *L. subcarnosus*, blue. *Mathiola bicornis*, the Night-scented stock, has no beauty, but is highly prized for its delicious perfume in the later hours of the day. *Mignonette*, the large-flowering and the red-flowering. belong to the second row, being more bushy and taller than the common mignonette. *Nemophila maculata*, white and purple. *Œnothera tetraptera*, white. *Papaver Rhœas nanus*, dwarf double poppy, very showy for second row. *Silene orientalis*, rose. *Specularia speculum*, Venus's Looking-glass, lilac and white. *Statice spicata*, rosy pink. *Omphalodes linifolia*, Venus's navelwort, blue. *Viscaria oculata*, rose; *V. elegans picta*, carmine. *Whitlavia grandiflora*, blue.

THIRD ROW.—(2 to 2½ feet).—*Amaranthus caudatus*, Lovelies-bleeding, red; *Calliopsis Burridgi*, crimson, fine; *Calan-*

drinia grandiflora, purple; *C. Drummondi*, yellow and brown; *Campanula attica*, violet; *C. media*, white, rose, and blue: there are seven distinct and fine varieties. *Centaurea cyanus*, Cornflower, various; *C. depressa*, blue; *C. moschata*, Sweet Sultan, various; *Chrysanthemum Burridgeanum*, yellow and crimson; *C. atrococcineum*, crimson; *C. purpureum*, purple: a fine group. *Clarkia elegans*, lilac; *C. elegans plena*, double rose. *Delphinium ajacis*, tall Rocket Larkspur, various: *D. chinenis*, a splendid blue-flowering annual or biennial. *Erysimum Peroffskianum*, orange, fine. *Godetia rosea*, rose; *G. Whitneyi*, crimson. *Helichrysum bracteatum*, various. *H. brachyrhynchium*, yellow and brown. *Hibiscus Africanus*, straw and brown. *Lupinus roseus*, rose. *Mimulus Tilingi*, fine yellow. *Œnothera bistorta Veitchiana*, yellow and crimson. *Papaver rhœas*, French and Pæony-flowered Poppy, various and very showy, but soon over. *Salpiglossis coccinea*, scarlet; *S. variabilis*, various. *Waitzia aurea*, gold. *Xeranthemum annuum*, various.

FOURTH ROW (3 to 5 feet).—*Agros-

GODETIA WHITNEYI.

temma coronaria*, rose. *Amaranthus cruentus*, purple red; *A. speciosus*, deep crimson; *A. hypochondriacus*, the Prince's Feather, purplish red. *Calliopsis bicolor*, yellow and brown; *C. lanceolata*, yellow. *Digitalis purpurea*, Foxglove, purple; there are several fine varieties, that named *gloxiniæflora* perhaps the best: it is figured at page 193. *Gypsophila gigantea*, white, useful for bouquets. *Helianthus argyrophyllus*, yellow. *Malope grandiflora*, crimson; *M. grandiflora alba*, white.

Fifth Row (6 feet and upwards).—*Helianthus grandiflorus*, yellow; *H. macrophyllus*, yellow; *H. Californicus*, yellow; *H. orgyalis*, yellow : four noble sunflowers well adapted for half-wild sunny places, but not for the best border. *Heracleum giganteum*, the giant cow-parsnip, a magnificent biennial for a half-wild spot, especially in damp soil. *Ipomæa purpurea*, the major convolvulus, one of the loveliest of twiners, various, ten feet. *Lathyrus odoratus*, the sweet-pea, is quite hardy and may be sown with other annuals in the open border in February. Amongst many fine varieties, the "Invincible Scarlet" must be named as one of the best, six feet. *Loasa aurantiaca*, orange, six feet.

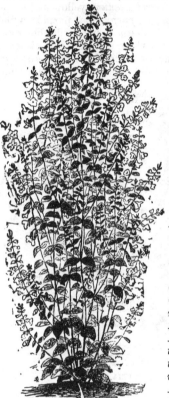

MIMULUS ILINGI.

Hardy Annuals best adapted for sowing in August. — *Calliopsis tinctoria, C. Atkinsoniana, Centaurea cyanus, Cladanthus Arabicus, Clarkia elegans, C. pulchella, Collinsia bicolor, C. verna, Convolvulus tricolor, Delphinium ajacis, D. consolida, Erysimum Peroffskianum, Eschscholtzia Californica, E. tenuifolia, Gilia tricolor, Godetia rubicunda, G. lepida, Iberis umbellata, Leptosiphon luteus, Limnanthes Douglasi, Malcomia maritima, Nemophila insignis, N. maculata, Platystemon Californicum, Saponaria calabrica, Silene pendula, Whitlavia grandiflora.*

Annuals that bear transplanting well may be advantageously grown for the purpose in turf-pits, the seeds being sown in August on a shallow bed of poor soil on a hard bottom.

Autumn-sown stocks, nemophilas, silenes, collinsias, erysi-
mums, and clarkias are particularly useful when raised in this
way and kept through winter, with no more protection than
just suffices to preserve them from injury by frost. The best
form of turf-pit is here figured, and the necessary directions
for its construction may be given in few words. The materials required are some good larch poles, some rough planking,
a good stock of turf, and a sufficient number of strong frame-
lights or sashes of a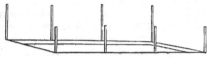
proper size. Mark
out the place for the
pit, choosing a dry
slope facing the south,
if possible, for damp is a greater enemy than frost to all
unheated structures. For a substantial working pit of good
capacity, the following inside measurements are here recommended:—Twelve feet long, five feet wide, three feet
deep at the back, two feet in the front. Having marked
out the ground, dig it out to a depth of twelve inches,
so that the inside of the pit will be that depth below the
level of the ground outside; then drive in short poles
at the four corners, and attach a rough plank along the edge
of the excavation all round, against which to lay the first
layer of turves. Then dividing the twelve feet space into
three equal parts, drive in four other stout poles for the sash
pieces to rest on, and then begin to pile the turves. These are
to form four solid walls to be laid down level with the ground
outside, neatly built up, beginning by laying them close to the
rough planking round the pit till level with the top of the
poles. If the walls are six inches thick they have sufficient
solidity, but they may be eight or nine inches with advantage.
 When these are
completed, trim
them off neatly
where they re-
quire it, observing
that the summit
should slope a little downward, to throw off rain and
prevent any trickling into the pit; and also let the outside be as regular as possible, that wet may not lodge
anywhere. A labourer accustomed to the handling of turf
would complete this part of the job in a few hours, and

finish it off as neatly as a brick-built wall. Then, for the sashes to rest on, nail a strip of board of sufficient width to lap over the turf to carry off rain; and fit three of the ordinary three and a half feet sashes, well painted and glazed, and your pit is complete.

To complete the pit for the reception of plants, make a bed of clean-sifted coal-ashes inside to plunge the pots in, or lay down a bed of brick rubbish, and on that one foot depth of sandy loam for the plants. During severe weather thatched hurdles would be the most useful covering to assist in keeping out frost. Pits of this kind are not only valuable in winter for preservative purposes, but in spring, when cleared out, they are useful for raising annuals and early vegetables for planting out. Two feet well-worked dung, with six inches of mould on the top, would make hotbeds of them at once: and during the whole year round, they could be kept in active use, and if well made at first would last a lifetime.

GARDEN GATE IN CHINESE STYLE.

CHAPTER X.

THE ROSE GARDEN.

IT is necessary to the completeness of this work that it should contain at least one chapter on the cultivation of the rose. But it is not necessary that any elaborate disquisition should be attempted; for although the subject invites us to be diffuse, and is known to be exhaustless, very much of useful information may be conveyed in a few words, and it is part of an author's duty to take quick and comprehensive views of things in the preparation of a small volume on a large subject.

We will first suppose that the reader has a garden of some extent, and would gladly institute a feature in the form of a rosarium, or compartment devoted exclusively to roses. We begin by presenting a plan for the purpose, which may be adopted in its integrity, or modified by a little careful manipulation. It must be understood, however, that this is not a fancy sketch, that may be altered *ad lib.*, as a mere design on paper; it is a plan of a rosarium that we have ourselves formed and planted, and found sufficient for our own enjoyment, and the satisfaction of a few critical friends who are known to be half mad on the subject of roses. The plan is drawn on a scale of twenty-four feet to one inch, and, if carried out on that scale, would require an oblong plot of land measuring about 140 feet in length by 90 to 100 feet in breadth. It consists of an oval occupied with grass and roses, enclosed with a parallelogram of hornbeam, or clipped yew, or of mixed plantation. In the centre is a basin of water fifteen feet in diameter: this is enclosed with a low fence of common China roses, very carefully trained. A fountain might be appropriately introduced here. The walk round the basin opens into four main cross walks five feet wide, and four secondary walks three feet wide, which communicate with the oval walk within the boundary of the trellis, four feet

wide. The spaces between these walks are filled in with
grass turf, in the four largest compartments of which are
small horseshoe-shaped beds, filled with dwarf China roses.
Four suitable sorts would be, *Belle de Florence, Eugene Beau-
harnais, Henry V.*, and *Napoleon.* The rustic trellis is clothed
with free-growing Perpetual, Noisette, and Tea-scented roses
in variety, affording space for about forty plants. The ten-
feet wide belt between the trellis and the outer elliptical
boundary walk is embellished with narrow scroll beds filled
with dark dwarf China roses in distinct masses of colour on
a flat groundwork of light China roses of only four sorts, one
sort in each compartment. For the eight scrolls, the following
sorts would be suitable, one sort in each scroll, namely, *Abbe
Mioland, Cramoisie Eblouissante, Cramoisie Superieur, Fabvier,
Henry V., Marjolin de Luxembourg, President d'Olbecque, Prince
Charles.* For the four small shield-shaped beds between the
scrolls the little Noisette *Fellenberg* would be suitable, or the
crimson miniature China rose. Four sorts of light China roses
will be required to fill in the groundwork, and there could not
be a better selection than *Mrs. Bosanquet, Aimee Plantier,
Alexina,* and *Madame Bureau.*

Beyond the outer elliptical walk are four spaces filled with
grass turf, forming the corners within the shrubbery boundaries. In each of these is a bed six feet wide, in the shape
of a letter L, affording room in each for one centre row of
mixed standard roses between two rows of mixed bush roses,
all of them Hybrid Perpetuals. The ovals are filled with
mixed bush roses. In each of the four compartments there
are three specimen trees, which may be conifers, but it would
be preferable to adopt standard weeping Ayrshire roses, and
gigantic bushes of Alba roses to form distinct and striking
features. Instead of a boundary of hornbeam or yew, a
palisade of climbing roses, or a plantation of bushes and
standards mixed might be adopted.

The question will occur where should such a garden be
formed, within view of the windows, or far away? We reply,
"far away;" for a rose garden should be in its season a wonder
to be sought, as, when its season is past, it is a wilderness to
be avoided, except by the earnest cultivator, who will never
cease to bestow on his roses all the care they require, asking
his friends to admire them, and share with him the joy of
their blooming when they are at their very best, and vindicate

DESIGN FOR ROSE GARDEN.

in their own magical, bountiful way, the devotion of their owner to their ways and wants.

The most simple form of rosarium will suit the majority of our readers, no doubt, and we advise those who do not require an elaborate plan to form on a part of the lawn farthest removed, and if possible hidden from the house, a few large beds in a group, and plant them with Perpetual roses, always giving the preference to dwarf bushes rather than to standards; because, generally speaking, bushes (especially if on their own roots), thrive in a more satisfactory manner than standards, and produce a far greater profusion of flowers. But here we must face the grave question, *why* one form of rose should be preferable to another, and we must attempt an answer in order to start the beginner in rose-growing fairly on the road to success.

There are many modes of multiplying the roses; but for all general purposes we need only notice three of them. The standard roses commonly met with, are obtained by inserting buds of named roses on the young shoots of English briers in the month of July. The operation is called "budding," and constitutes an important mystery of the rose craft. Bush roses are obtained by the budding process; but an Italian brier, known as the Manetti rose, is employed for the purpose. It is a free-growing, very free-rooting, bluish-leaved brier not adapted to form standards, but well suited for bush roses if the buds are inserted very low down, in fact immediately over the roots of the briers, so that when they grow they will spring as it were from the ground, instead of from the stems in which they are inserted. Both bush and standard roses may be obtained on their own roots by striking cuttings or buds, or making layers of named roses, and the month of July is the best season in all the year for these operations.

BRIER ROSES are admirably adapted for deep loamy and heavy clay soils. In any and every case the ground intended to be planted with roses should be well drained, and if the subsoil is anything approximating to a clay or deep rich loam, brier roses may be planted with a fair prospect of success. To make brier roses is a simple matter enough, when you know how, but very mysterious short of that point. In the "Rose Book" ample instructions are given for the multiplication of roses in all possible ways; but here we must cut the matter short by saying that the art of budding may be learnt in five minutes on the ground with the help of the demonstrations and explanations of one who is somewhat expert in performing it, but will be very slowly apprehended by the best written instructions, however freely illustrated and "adapted to the meanest capacity."

MANETTI ROSES are adapted for all soils and situations; but have an especial value for gravelly, chalky, and worn-out soils, because of the abundance of roots the Manetti brier produces, and its consequent power of obtaining nourishment in comparatively barren lands. When this stock is employed for dwarf roses (and it is not suited for the production of standards) the stems should be budded near the ground; indeed a little of the earth should be removed to enable the operator to insert the buds as low down as it is possible to find a green lifting bark on which he can operate with a hope of success.

OWN-ROOT ROSES are, generally speaking, the most valuable of all. They are such as have roots of their own, that is to say, they are not obtained by budding or grafting, but by the striking of buds or cuttings, or putting down layers; in each case the rose makes roots for its own sustenance, instead of being made to depend on the roots of briers, manettis, or any other stocks. Any one who has had a little experience in the propagation of bedding plants ought to find it easy and agreeable work to produce a stock of own-root roses. There are many modes of procedure open to the choice of the proficient. The simplest of all methods may be described in a few words:—

There will be found on all the rose-trees in the middle of July, a number of plump, young, green shoots of the same year. As the seasons vary, so will the time vary for taking cuttings; and the best rule that can be given is, that they should be taken when about half ripe, the wood being still green but firm, for so long as it is decidedly soft and sappy it is unfit. The selected shoots should be cut up into lengths of about four inches each, and the lowest leaf should be removed. The soft tops of shoots should either be cut off and thrown away, or should be carefully struck in the same way that soft bedding plants are, in pans filled with sand, in a rather strong, moist heat. But the cuttings we have especially in view, consisting (say) of young wood as thick as a goose-quill, in lengths of four to six joints each, the lowest leaf only removed, will not require heat, but will quickly make root if planted

thickly in a bed of sandy soil, or even in a bed of cocoanut-
fibre, and kept close and moist, without ever being very wet,
or in a hot, stifling atmosphere.

To make own-root roses from buds is not quite so easy as
to make them from cuttings. The first step is to obtain a lot
of precisely the same sort of buds as would be required for
budding briers. The next thing is to prepare them in the
same way, without removing the wood or the leaves. The
wood, indeed, may be removed, but it is waste of time to
remove it; but if the leaf is removed the bud will simply die.
Having secured buds cut in the fashion of shields, without
removing the wood, and, above all things, without removing
the leaf that each must have when cut, plant these buds firmly
in pans filled with sand, or on a bed of light loam covered
with sand over a mass of fermenting material, or in a common
frame. All the leaves must stand up and be kept fresh by
frequent sprinkling, but there must be no slopping of water
amongst the buds, or they will rot: in fact, any excess of
moisture will ruin the best planned project for propagating
roses with equal certainty and rapidity with the total aban-
donment of the cuttings or buds to drought, by an act of for-
getfulness or intentional rose-murder.

To propagate by layers is the easiest plan of all; but it is
impossible to make many roses in this way, because two or
three are the utmost number obtainable from a shoot, whereas
by cuttings or buds a strong shoot will furnish material for
from twelve to twenty plants. But certainty may well compen-
sate for lack of quantity with many readers; and our advice
to lovers of roses who cannot see their way clear to strike
cuttings, is to make layers of them in July and August in
precisely the same way as carnations and picotees are layered.
Lastly, but not leastly: If you will wait until the middle of
September, you may then begin to multiply roses by what we
have designated "the currant-tree system." To make short
work of the subject, we may remark that roses may be struck
from cuttings precisely as currant-trees are struck; but the
business should be attended to while the roses yet have green
leaves upon them. Many try this system and fail. It is all
their own fault, for they allow the proper season to pass by,
and suddenly make a rush at the propagating when the
season for the work is past. From the middle of September
to the end of October is the proper time for the practice of

the currant-tree system of multiplying roses, and if the work is well done then, eighty per cent. of the cuttings will root. People who are blessed with a spirit of patience and perseverance may continue, or begin, to put in cuttings of roses in the open ground or in frames all through the winter months, say from November to February, and in favourable seasons may be wonderfully successful. But the risk of loss is great, and the only argument in favour of winter propagation is, that in peculiarly sheltered spots, where an early bloom is desired, winter pruning must be practised, and the prunings may be turned to account to make stock, provided only that nature will assist the enterprise. In the attempt to strike cuttings after the turn of the year, a cold frame and a bed of cocoanut-fibre and sand will be immensely serviceable. If the steady bottom-heat of a propagating house can be secured, first lay the cutting in a horizontal position, just covered with tan or fibre, in a warm, moist place for a week or so, to promote the formation of the "callus," and then insert them upright in sandy stuff in a temperture of about 50°, a few degrees more or less being of no consequence, provided only that the bed is neither burning hot nor freezing cold.

THE CHOICE OF ROSES.—For the decoration of the garden the course of procedure should not be the same as when roses are grown for exhibition. Elegance in the plant, and abundance of flowers, are the principal desiderata of garden roses. We will consider the mechanical part first, and then the floral. You will want standards and dwarfs to make a good plantation, and in reckoning the quantity required, it will be well to allow an average of two feet apart every way, and there should be three or four dwarfs to every standard. If it is but a small bed, your standards must not in any case exceed four feet in height, and better if only three feet. But with every increase of size in the plantation, you may take taller and taller standards, planting them in the order of their heights from front to back, if the bed has a front and a back; but from outside to the centre they should rise in height, if the bed is to be viewed from all sides. They must be planted at distances consistent with their several habits of growth—the strong-growing kinds two and a-half to five feet apart, and the weak ones, such as we use for edging (say Fabvier and its kin), eighteen inches apart. The effect to aim at is a close rich mass of bush roses, with standards rising up out of the mass in successive ranges of altitude, these standards being rather sparely sprinkled, too few being a better rule than too many. You have but to select suitable sorts, and prepare for them as here advised, and you shall have a glorious display, and your standards will improve equally with your bushes, even if you never give them one drop of water. You know what our Lord said of the man who built his house upon the sand. That applies not alone to the spiritual life of man, but to every one of his worldly adventures. Begin well, make the foundation safe, and you may hope to prosper. Stick a rose-tree in a hole, and expect it to die; plant it properly, and it will pay you for your pains.

Now, as to the floral qualities. For garden roses we need vigorous growers that are sure to flower freely, and that will, for a considerable length of time, contribute to the gaiety of the garden. The best garden rose is one that an exhibitor would despise: it is a Bourbon, named *Lord Nelson*. One of the worst, is one that exhibitors used to make much of: it is a Hybrid Perpetual, named *Louis XIV*. There is something in making a good selection, depend upon it. We shall not enumerate all the good roses for a garden display, but will undertake that in this selection there shall not be a bad one.

A SELECTION OF GARDEN ROSES.

STANDARDS.—*Gloire de Dijon, Marechal Niel, Aimée Vibert, La Biche, Ophirie, Triomphe de Rennes, Bourbon Queen, Mere de St. Louis, Louise d'Arzens, Souvenir de la Malmaison, Anna Alexieff.*

VIGOROUS BUSHES.—*Baronne Prevost, Charles Margottin, Coquette des Alpes, Deuil de Prince Albert, Elizabeth Vigneron, Eugene Appert, General Jacqueminot, Gloire de Ducher, Glory of Waltham, Imperatrice Charlotte, Jules Margottin, La Rhone, Lord Nelson, Lord Palmerston, Madame de Cambaceres, Madame Domage, Madame Knorr, Mrs. John Berners, Pius IX., Sir Joseph Paxton, Souvenir de Ponsard, Triomphe de Caen.*

DWARFEST FOR FRONT LINES.—*Fellenberg, Fabvier, Mrs. Bosanquet, Madame Bureau, Common China.*

ROSES PEGGED DOWN.—One of the most effective and interesting modes of producing a fine display of roses is to grow them on their own roots and the branches pegged down. The management is very simple. A deep, rather rich,

HYBRID PERPETUAL ROSE.

(Globular form.)

loamy soil is necessary, and the position selected for the bed should be rather open, but not exposed to rough winds. As a rule, where roses can be grown as standards or bushes, they can be grown pegged down. The most important point is to plant only those kinds which are known to do well, and are on their own roots. The best of the large number of varieties for this mode of culture, are *William Griffiths*, *William Jesse*, *Charles Lefebvre*, *Anna Alexieff*, *Senateur Vaisse*, *Alfred Colomb*, *Baronne Prevost*, *General Washington*, *General Jacqueminot*, *Jean Goujon*, *La Reine*, and *John Hopper*. This list, short though it is, contains the cream of all the best varieties suitable for pegging down. Do not plant worked roses, for if they are worked upon any stock, the suckers from the stock will, sooner or later, come up and overgrow the roses unless a continual warfare is waged against them.

It does not matter much when they are planted out of pots, but the most favourable months are October and April. Whether planted in autumn or spring, let the surface of the bed be well trodden soon after the planting is completed, as the rose under all circumstances prefers a firm soil. Also tread the surface every spring after the plants are pruned and the beds forked over. If the plants are not more than a foot high when planted, do not prune the first year, but simply peg down the strongest growth. On the other hand, if they are strong, with shoots two feet long and upwards, just take off about eight inches of the points in March, as it is not desirable to let the young plants have much old wood to support the first season. Aim at a vigorous growth, so as to have plenty of flowering wood for the next year. The careful cultivator will take care that the plants do not want for water the first summer after planting, but after that time they are able to take care of themselves. Those who desire to have fine blooms throughout the season, must cut off a few inches of the flowering wood as soon as the first bloom is over, and give the beds a thorough soaking of manure or sewage water, every third or fourth day, for a short time. After the application of the manure water, the plants will soon start into a new growth, and furnish a supply of flowers, if the weather is mild, until Christmas. They will be well established in the soil in twelve months after planting, and will grow away vigorously, sending up strong shoots three or four feet in height. These must be pegged down in the

14

vacant spaces during the summer, to give the beds a neat appearance. If more young shoots are made than are likely to be wanted to cover the surface of the bed at a distance of nine inches apart, cut away the weakest at once, but do not peg down the young shoots until obliged to do so.   If done before the end of August, the lower buds will most probably start into growth.

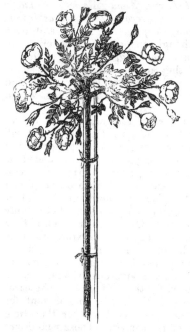

There must be a fresh supply of young wood every year, and the old wood cut away to make room for it. A greater quantity and finer blooms are obtained from young wood than from old.

Early in December cut away the old wood, and take away the pegs from the young growth, which has been pegged down so as to allow it to rise up a little from the ground. If kept closely pegged to the surface, the shoots are in mild winters influenced by the warmth of the earth, and start into growth early, and the risk is incurred of injury by spring frost. After the pruning is completed, cover the surface of the bed with

STANDARD ROSE TREE.

two or three inches of fat manure, and let it remain.

Early in March cut back all the shoots according to their strength, the strongest to two feet and the weakest to twelve or eighteen inches. This is quite long enough for plants that are not more than two feet apart each way. The manure which was put on in the winter, should then be forked in and the beds well trodden as advised above.

Thus far we have considered how to obtain roses in quantity, and now quality must engage our attention. There are certain principles common to rose culture in all the phases and aspects of the art. Thus the rose loves a good, deep, rich loam, plenty of moisture all the summer long, and a sur-face-dressing of manure every winter. All these leading principles must be observed in rose-growing for quality; but at a certain point this branch of the subject separates from all the rest, and we come abruptly on certain excep-tional circumstances, as will be seen by the very next para-graph.

EXHIBITION ROSES. —The brier has been condemned by the writer of this again and again, as utterly unfit for common use in the produc-tion of garden roses. To speak of it truth-fully and collectively, it may be said to be the curse of the su-burban garden, for in all small gardens we see ghastly spectres

HALF STANDARD ROSE TREE.

called "standard roses;" the best of them are mops, the worst are scarecrows. Now, good rosarians, bear in mind from the very first, that the brier is your best friend; and if you go into rose-growing in earnest, you will not make much progress until you master the brier root and branch, and know every pulse that throbs in the "secret chambers of its blood." Yes, the brier is a grand agent in the production of quality roses, and you must learn how to plant it, how to bud it, how to transplant, keep, and improve it, and, at last, how to cut from it the rose that shall bring you golden honours in reward for all your toil. Write it down among things not generally known, that brier roses one year budded will, if grown vigorously, and severely thinned, furnish (generally speaking) finer individual blooms than any other form of roses, and to these young briers you should look for your supplies from year to year of roses for the exhibition table. Write down, again, that to keep up the stock of the right sort of plants, you must plant young briers every autumn, and bud those young briers every summer. Write it down again, that, whereas garden roses should never be hard-pruned, but allowed to grow freely, and bloom plentifully ; hard-pruning and severe thinning-out of flower-buds must be practised in the quarter where the show roses are grown. Lastly, to quit these sweeping generalities, write it down, that every separate variety has its own peculiar habit of growth, and the Brier does not suit all alike, but some thrive better on the Manetti, and others better on their Own Roots ; and it is well to try every variety, every way, and have as many strings to your bow as you can see and handle with safety. The figures will indicate, in the obvious distinction between old and young wood, how roses should be pruned for the production of fine flowers.

In our practice, budding has well-nigh changed into summer-grafting, as told ten years ago in the " Floral World," and set forth at length in the " Rose Book." We bud and bud as usual, if the bark rises nicely ; but if, as will happen in a season of drought, the bark does not rise nicely, and the core does not jump out and leave a clean shield, we leave the core alone, and make a true graft of what would have been, in a sap-running season, a genuine bud. This is a great stride in practice, which perhaps only the advanced practitioner will thoroughly understand, for it insures that all the briers to be budded will be budded, and that, too, as suits our own convenience, without standing still for rain and making a rush at the work the moment a black cloud comes in sight. You may see something in this to suit the philosophic humour, and mayhap, may con over, while engaged in the delightful task of marrying

" A gentler scion to the wildest stock,"

those lines of Spenser, in which he tells the secret of success
in every pursuit and aim of life :—

> " In vain do men
> The heavens of their fortune's fault accuse,
> Sith they know best what is the best for them ;
> For they to each such fortune to diffuse
> As they do know each can most aptly use.
> For not that which men covet most is best,
> Nor that thing worst which men do most refuse ;
> But fittest is, that all contented rest
> With that they hold : each hath his fortune in his breast."

In selecting roses for quality, you must know in what
quality consists. You will soon discover that there are many
kinds of roses of high quality which revolve, as it were, like
planets around the central Ideal Rose that grows only in
the innermost chamber of the Amateur's cerebrum. Every
individual rose, when you come to play critic, will be found to
have its own defects as well as its own excellences, and there
will be a separate reason, therefore, in every separate case for
growing that particular rose. To enter into a discussion of
properties, however, would carry us far beyond reasonable
limits, and therefore we make short work of this part of the
subject by a series of brief explanations of the terms employed in the description of exhibition roses.

The leading contours of roses may be comprised under
the following heads : Globular, Reflexed, Expanded, Cupped,
and Half-cupped, or Tazza-shaped ; which forms will result
from the centre, the face, the profile, and the size, shade, and
depth of petal.

1st, The Globular is almost invariably double, and well-

formed, the centres being full of
small leaflets, the exterior petals
sometimes folded over at the
points with great elegance and
regularity, as in Chabrilland and
Senateur Vaisse. A variety of

REFLEXED.
GLOBULAR.　this form sometimes arises from
the whole of the petals being incurved as in Madame Pierson
and Robert Fortune. The petals are large, and profile
deep.

2nd. The Reflexed has a high centre, from which the

petals turn ever, increasing in size to the outer row; they are sometimes imbricated, as in Madame Vidot; profile deep.

3rd. The Expanded, is a modification of the last, being larger in diameter, flatter at the centre, and with a much shallower profile.

EXPANDED.

4th. The Cupped is not always full. The face is flat, and the outer petals large, holding up, as it were, the interior; profile deep.

5th. The Half-cupped or Tazza-shaped is more expanded than the last, fuller in centre, and larger in diameter; it is in these last two classes that deficiencies in doubleness most frequently occur.

CUPPED.

TAZZA SHAPED.

This remark, however, does not apply to the Bourbons, which for the most part have excellent centres. The face is flattish, and the profile somewhat shallow.

A SELECTION OF THE FINEST EXHIBITION ROSES.

H. P. means hybrid perpetual; N., noisette; T., tea. *All in the list are of extra size.*

NAME.	CLASS.	RAISER.	DATE.	DESCRIPTIVE REMARKS.
Louis Van Houtte	H. P.	Lacharme	1869-70	crim. shad. black
Clemence Raoux	,,	Granger	,,	light rose
Princess Christian	,,	W. Paul	,,	rosy peach
Comtesse d' Oxford	,,	Guillot père	,,	bright rosy tint
Edouard Morren	,,	Granger	1868-69	rose
Julie Touvais	,,	Touvais	,,	rose
Marquise de Montemart	,,	Liabaud	,,	tinted white
Nardy Frères	,,	Ducher	,,	darkish rose, large
Madame Jacquier	,,	Guillot fils	,,	purple and crim.
Miss Ingram	,,	Ingram	,,	pale pinkish tinge
Duke of Edinburgh	,,	Paul and Son	,,	dark crim. scarlet
Victor de Bihan	,,	Guillot père	,,	rich rose
Thyra Hammerich	,,	Victor Verdier	,,	pale rose
Reine Blanche	,,	Damaizin	,,	tinted white
Montplaisir	T.	Ducher	,,	yellow and buff
Baroness Rothschild	H. P.	Pernet	1867-68	rose shaded white
Elie Morel	,,	Liabaud	,,	rose
La France	,,	Guillot fils	,,	rose shaded white
Madame Noman	,,	Guillot père	,,	tinted white
Pitord	,,	Lacharme	,,	deep red scarlet
Black Prince	,,	W. Paul	,,	very dark
Horace Vernet	,,	Guillot fils	,,	crimson scarlet
Annie Wood	,,	E. Verdier	,,	,,

NAME.	CLASS.	RAISER.	DATE.	DESCRIPTIVE REMARKS.
Monsieur Noman	H. P.	Guillot père	1867-68	shaded rose
Prin. Mary of Cambridge	„	Paul and Son	„	pale rose
Felix Genero	„	Damaizin	„	light crimson
Abel Grand	„	Damaizin	1865-66	silvery rose
Alfred Colomb	„	Lacharme	„	fiery red
Jean Lambert	„	E. Verdier	„	scarlet crimson
Marie Rady	„	Fontaine père	„	rose
Dr. Andry	„	E. Verdier	1864-65	carmine crimson
Duchesse de Caylus ...	„	...	„	brilliant carmine
Marguerite de St.Amand	„	Amand	„	rosy flesh, superb
Xavier Olibo	„	Lacharme	„	crimson and black
Marechal Niel	Tea or Noisette }	Pradel	„	golden yellow, superb
Alpaide de Rotalier ...	H. P.	Campy	1863-64	pale rose
La Duchesse de Morny	„	E. Verdier	„	bright rose
Lord Macaulay	„	W. Paul	„	crim.shaded black
Madame Victor Verdier	„	E. Verdier	„	carmine crimson
Marie Baumann.........	„	Baumann	„	brilliant red
Pierre Notting	„	Portemer	„	purple red, deep
Beauty of Waltham ...	„	W. Paul	1862-63	cerise
John Hopper	„	Ward	„	bright rose
Charles Lefebvre	„	Lacharme	1861-62	crim. shad. purple
Madame C. Joigneaux	„	Liabaud	„	large bright rose
„ *C. Wood* ...	„	E. Verdier	„	vinous crimson
Maurice Bernardin ...	„	V. Verdier	„	deep scarlet
Prince C. de Rohan ...	„	E. Verdier	„	crim. shad. black
Vicomte Vigier	„	V. Verdier	„	purple crimson
Madame Furtado	„	V. Verdier	1860-61	rich rose
Senateur Vaisse	„	Guillot père	„	crimson scarlet
Victor Verdier	„	Lacharme	„	brilliant rose

" How much of memory dwells amidst thy bloom,
    Rose ! ever wearing beauty for thy dower ;
The bridal day, the festival, the tomb,
    Thou hast thy part in each, thou stateliest flower.
Therefore with thy sweet breath come floating by
    A thousand images of love and grief ;
Dreams filled with tokens of mortality,
    Deep thoughts of all things beautiful and brief.
Not such thy spells o'er those that hailed thee first,
    In the clear light of Eden's golden day,
There, thy rich leaves to crimson glory burst,
    Linked with no dim remembrance of decay."

CHAPTER XI.

THE AMERICAN GARDEN.

THE money spent on rhododendrons during twenty years in
this country would nearly suffice to pay off the National Debt.
If the reader fails to apprehend the force of this remark, its
meaning may be reached through an inspection of all the
villa gardens of the country. In these villa gardens, that is
to say in a considerable proportion of them, will be found
numbers of perishing rhododendrons inhabiting common
borders, and associated with laurels, aucubas, hollies, and
such like; all these other inmates of the borders being
perhaps in a thriving state, while the rhododendrons are
going, going, from their pristine buxom beauty to a condition
of shrunken starvedness that tells to sage beholders that their
death is near. The very bad practice of planting standard
roses on grass turf has its parallel in the equally bad practice
of planting rhododendrons in the common border, where,
unless by some peculiar accident the soil happens to suit
them, they must die, and in the process of dying become
hideous long-armed things that no one would wish to save.
It is a most important part of our business to warn the reader
against waste of money, and time, and hope, in ill-advised
adventures in the garden. We therefore protest against the
wasteful and ridiculous practice of treating the rhododendron
as adapted for any and every position in which a handsome
evergreen shrub may be required. No matter how cheap,
how common, or how hardy, this noble plant is peculiar in its
requirements, and must be humoured, or it will dwindle and
die. Deal with it aright, and it grows rapidly, flowers freely,
and becomes one of the grandest ornaments of the garden ;
but begin with it in the wrong way, and it can only serve as an
evidence that somebody near at hand is not yet quite accomplished in the art of gardening.

The rhododendron represents a group of hardy shrubs of the same botanical family, and immediate relatives of the erica, all of which require a soil containing the least possible proportion of the carbonates or salts of lime, and the greatest possible quantity of humus and siliceous grit. The best soil for the whole family is a peat containing much sand and much vegetable fibre; but we are not restricted to peat, and it is by no means a difficult matter to prepare soil for the shrubs of this class, in districts where peat is not to be found. It must be understood that what are called "American plants," have extremely fine hair-like roots, which are quite incapable of penetrating a harsh soil such as clay or heavy loam. Therefore in a harsh soil they die almost as surely as if planted in chalk or gypsum, which are poisons to them; but in mixtures of sand, leaf-mould, and fine loam they will thrive, and almost any clean pulverized product of vegetable decay may be added to a sandy staple to prepare it for them. Nay, even stable manure, thoroughly rotted to powder, may be employed as one ingredient in a mixture, though it is regarded as equally injurious with lime and chalk, which it certainly is not if old and powdery, and employed in proportion not more than a sixth of the whole bulk. A simple and cheap method of preparing a substitute for peat may be resorted to in cases where old upland pastures or common lands are being broken up. The workmen should be required to take off, not the top spit, nor even the ordinary thickness of a turf, but *a skin* consisting almost wholly of the grass and its roots, the texture of which should resemble thick felt, or carpeting. The skinning process is accomplished by means of the broad end of the pick, by a chipping process, with great rapidity after a little practice. If this skin is laid in small heaps in the full sun, the grass withers instantly, and the stuff may be used at once. We have, indeed, used it as fast as it was removed, having it chopped into pieces of the size of one's hand, and mixed with one-third of its bulk of sharp sand or sandy gravel, and have at once planted rhododendrons in the mixture, and had the satisfaction of seeing them years afterwards grown to lusty giants, and renowned in the district for their glorious appearance when in flower. In any case, whether the soil on the spot will be sufficient, or a mixture is prepared or peat purchased, a depth of not less than two feet must be provided, and it will be better practice to make the beds four feet deep, not because the roots of American plants penetrate deeply, for, in truth, they do not, but to secure a constantly moist under-crust, for shallow beds of peat or sandy mixtures become as dry as so much pebbles during the heat of the summer, and the result is, serious injury to the American plants.

All these plants bear partial shade well, but an open position is always to be preferred for them, because their flowering depends on the perfect ripening of the young wood. Shelter is desirable, and hence the American garden should, if possible, be formed on the south or south-west side of a shrubbery or woodland. We frequently meet with large clumps of rhododendrons in entrance-courts, and if they are but good of their kind, they are the most appropriate of all our hardy shrubs for the embellishment of such a spot. Their attitudes and solid dark green heads and their magnificent flowers constitute them proper for front court, terrace, and lawn shrubs, and in all those highly-dressed positions rhodo-dendrons should be chiefly adopted—that is, if American plants are employed at all—because of their orderly growth and massive appearance. The hardy Azaleas are deciduous, which in part unfits them for the front-court and terrace. The Kalmia is capricious, and should never be planted in any quantity or in any important position, until it has been tried and found to answer. A deep, gritty peat, abounding in vegetable fibre, and a position exposed to all weathers, are conditions favourable to the prosperity of the Kalmia, which is useless when planted in make-shift soils, or in smoky or shady localities. The Ericas love sand, and will thrive in mere gravel if deep enough; but any soil that suits the rhodo-dendron will answer for them, especially if they are planted in the first instance in a mixture of good sandy peat to give them a fair start. As for Andromedas, Ledums, and the rest of the beautiful and interesting plants of this family, all that need be said about them is, that they appear to better advantage when planted to form outside belts to masses of rhododendrons, azaleas, and kalmias, than when put into com-partments by themselves.

In planting American shrubs it is desirable to plant thick enough to produce an effect at once, not only because they are planted for effect, but also because they thrive better when they cover the ground and by their own leafage check evaporation. Now to do this need not be so costly a business

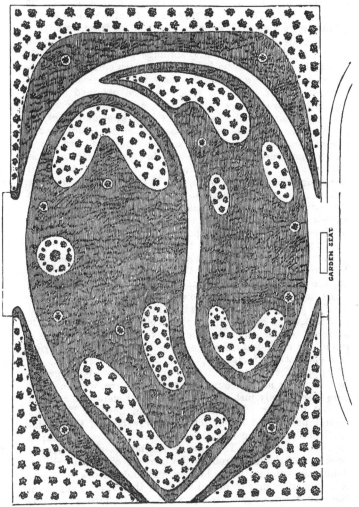

DESIGN FOR AN AMERICAN GARDEN.

GARDEN SEAT

as it will for the moment appear to the reader, who remembers
that some of the choicer kinds of rhododendrons are sold at
ten to thirty shillings for a very small plant, and large trees
of the more esteemed varieties are sold at from twenty to
fifty pounds each. Good unnamed seedling plants may be
obtained at from thirty to fifty shillings per dozen, and first-
rate named kinds at from half-a-crown to ten shillings a plant.
For thickening new plantations there is nothing better than
the common *R. ponticum*, which may be obtained at from
fifteen to fifty shillings per hundred, according to size. In
planting them in belts or beds, or long sweeping lines, no
great depth of plants is necessary to create bold, effects,
because, from the dense leafage and tendency to flower freely,
and the immense trusses of bloom they produce, three, or at
the most four, plants deep, when they have arrived to a good
flowering size, will be enough. When marking out the
positions, the planter ought to have a sufficient knowledge of
the varieties to foresee at what distances they may be planted,
so that they do not require moving and rearranging after four
or five years. In fact, the positions of the named varieties
ought to be permanent; and the cheaper kinds that are employed for filling should be removed as soon as ever they can
be spared. This is far better than planting all the good kinds
thick enough to form the beds at once, and then having to
move and rearrange the whole again in a few years.

The plan presented at page 219 is adapted to a spot an acre
in extent, but it may be (although less effective) reduced to half
an acre. If one acre could be spared, it would constitute a
grand and noble feature where its dimensions could be carried
into effect without cramping or disfiguring the outlines. It
may appear at first sight that the extent of grass is too great;
but, knowing how effective these subjects are, we are quite
sure there is only just enough greensward shown to create a
perfect harmony when the trees are in flower. The beds
should be planted chiefly with the very best of the rhododendrons, with here and there an azalea, and a few hardy
heaths for variety. The outside plantations should have a
back row of rhododendrons, and a mixture of azaleas, rhododendrons, kalmias, hardy heaths, and andromedas in front.
A few conifers may be placed singly on the grass, as shown on
the plan, to give lightness and variety and afford key-points
to the scene.

FIFTY SELECT HARDY RHODODENDRONS.

Alarm, Album elegans, Album grandiflorum, Archimedes, Atrosanguineum, Barclayanum, Blandyanum, Brayanum, Bylsianum, Charles Dickens, Concessum, Cruentum, Delicatissima, Elfrida, Everestianum, Fastuosum flore-pleno, Francis Dickson, Guido, Hogarth, John Spencer, John Waterer, Lady Armstrong, Lady Clermont, Lady Eleanor Cathcart, Lady Francis Crossley, Lord John Russell, Lucidum, Maculatum superbum, Magnum Bonum, Milnei, Minnie, Mrs. John Clutton, Mrs. John Waterer, Mrs. R. S. Holford, Mrs. William Bovill, Ne Plus Ultra, Nero, Perfection, Purity, Roseum elegans, Roseum grandiflorum, Standard of Flanders, Stella, Sherwoodianum, The Queen, The Warrior, Titian, Victoria, William Austin, William Downing.

A SELECTION OF HARDY AZALEAS.

Amœna, Ardens, Aurantia major, Bessie Holdaway, Calendulacea eximia, Coccinea major, Coccinea speciosa, Cuprea, Cuprea splendens, Elector, Elegans, Florentine, Fulgida, Gloria Mundi, Invictissima, Marie, Dorothé, Marie Verschaffelt, Mirabilis, Nancy Waterer, Prince Frederick, Speciosa atrosanguinea, Straminea, Sulphurea, Van Houtte, Viscocephala.

A SELECTION OF AMERICAN PLANTS FOR EDGING PURPOSES.

The following are eminently desirable for forming dwarf belts round beds occupied with hardy azaleas and rhododendrons :—*Andromeda floribunda, A. formosa, A. pulverulenta, Daphne cneorum majus, Erica cinerea alba grandiflora, E. c. atropurpurea, E. c. coccinea, E. c. rubra, E. herbacea carnea, E. vagans carnea, E. v. rubra, E. c. v. alba, Menziesia polifolia alba, M. p. atropurpurea, Kalmia angustifolia glauca, K. a. rubra, K. glauca, K. g. superba, K. latifolia major splendens, K. l. myrtifolia, Ledum buxifolium, L. thymifolium, Pernettya mucronata, Polygala chamæbuxus, Vaccinium frondosum venustum, V. ligustrifolium, Calluna vulgaris, C. v. coccinea, C. v. alba, C. v. aurea, C. v. flore-pleno.*

CHAPTER XII.

THE SUBTROPICAL GARDEN.

THE " subtropical garden," as at present understood in this country, is an importation from Paris, of limited, and indeed almost questionable value. Considered in close accordance with its designation, it requires us to expose to the common atmosphere, and to all possible changes of weather, any and every kind of stove plant that, owing to its distinctive outlines, or brilliant colours, may be considered suitable for purposes of outdoor embellishment. We may set apart a plot of ground for the purpose, and having crowded it with cannas, palms, tree-ferns, caladiums, begonias, and other elegant and valuable stove and greenhouse plants, pronounce the affair a subtropical garden. It may be a good or a bad example, as taste and judgment have or have not been employed in its production, and we may premise that, unless taste and judgment are employed, the subtropical garden is likely to prove the most ludicrous of all possible garden failures. In the first consideration of the subject, it will be well to adhere to the contracted signification of the term " subtropical," in order to arrive at something like a clear perception of its bearings on matters practical. We are required, say, to embellish a garden with plants many degrees more tender than our familiar geraniums, verbenas, and petunias, and we may reasonably expect to find the task increasingly troublesome and hazardous, in direct proportion to the tenderness of the subjects that it is proposed to employ. We may first reflect with advantage that this is a sub-arctic clime ; the summer is seldom in any sense established before Midsummer Day ; that for subtropical plants its duration extends to two months at the utmost ; and that night frosts have actually occurred, and have been registered both on thermometers and the face of nature, in the usually sunny months of June, July, and

August. It follows of necessity that the more we trust to tropical or subtropical plants for the embellishment of the flower-garden, the greater is the risk we incur that, instead of embellishing, we shall disfigure it. What the season is to be, we never know in time to make minute and special preparation for it, and a cold wet summer will damp our ardour, while it destroys the tender plants which in an hour of exceptional sunshine we may have drafted into the subtropical garden. The employment of tender plants for purposes of outdoor embellishment is a task that corresponds in degree of difficulty with the degree of tenderness of the plants employed. The nearer we go to the tropics for material, the nearer do we verge towards the impossible in the endeavour to adapt them to the average conditions of a British summer. In reference to some of the so-called subtropical plants, we may employ an Americanism to indicate the difficulty that attends their cultivation in the open air, by saying that it takes two men and a boy to hold one up; for, to speak the plain truth, many of the plants that we find in full fig in subtropical gardens are not worth, as they stand, a tenth part of the labour it has cost to place them there. A great musa torn to ribbons by angry winds, for example, may be likened in its sad fate to a magnificent passage of Shakespeare, which some noisy novice has just mouthed in costermonger tone to a "discerning public." But we are not to condemn subtropical gardening on account of the many men and boys required to hold it up, because it is a new thing, and we must always expect mistakes in the region of experiment, and in this case mistakes are scarcely less instructive than successes; for if the last teach us what to eat and drink in the way of vegetable beauty, the first will teach us what to avoid.

Gardening is always more or less a warfare against nature. It is true we go over to the "other side" for a few hints, but we might as well abandon our spades and pitchforks as pretend that nature is everything and art nothing. Our gardens are crowded with the plants of other climes, and these for the most part can only live so long as art supports them, for nature would soon kill them out, and plant their graves with chickweed, if left to her own sweet will. Therefore in subtropical gardening art is everything, and the artist must begin by preparing for his purposes an artificial soil and an artificial climate. Mr. Gibson has shown by his magnificent examples of subtropical gardening in Battersea and Hyde Parks, how to provide for the roots of subtropical plants a warmer soil and a more equable and genial climate than nature offers the artist for his work. By means of raised banks and mounds resting on a porous substratum of brick rubbish, the heat of the sun is caught and imprisoned in the soil; and by means of judiciously-disposed plantations of large-leaved deciduous trees, such as poplars, planes, and sycamores, the cold winds have their sharp edges blunted, and the protected region enjoys a more still and more genial atmosphere than the common world without, where the native flora takes its chance of storm or calm, of rain, and snow, and drought, of frost and sunshine. When we consider that subtropical plants require, in some cases, careful keeping in warm berths during winter, or careful raising from seed in spring, and in all cases careful preparation of the soil below and the air above for their tender nourishing when planted out, there can be no violence in pronouncing the general deduction that subtropical plants are not everybody's plants, and those who contemplate an indulgence of the new and extravagant fashion should count the cost and " cut the coat according to the cloth." Find the proper place and suitable means and talent equal to the task, and we shall be very grateful for an example of subtropical gardening,—first, because of a change away from the intensely strong flat colouring which has become obnoxiously popular; and, secondly, because we can better compare, and criticise, and enjoy the various characteristics of plants that are chosen principally for beauty of form: for in truth it is but seldom we can grasp the whole expression of a plant when we are restricted in our contemplation to a view of it under glass.

When we have made some progress in the artistic disposition of palms, ferns, and musas in the open ground, we shall not be slow to discover that many hardy plants may be associated with them to the advantage of artistic effect. Thus subtropical gardening always tends to subarctic gardening; for the true artist, to whom effect is everything and materials nothing, except as means to an end, will be always seeking for hardy plants equal in distinctive beauty, and all the interest that for the artistic eye belongs to form as apart from colour, to the most truly tropical or subtropical; and happy will be if some hedgerow fern or hemlock of the arctic marshes shall prove to be worthy of a place in the garden, in the same compartment in which tree-ferns, and palms, and castor-oil plants and wigandias are the proper subtropical oc- cupants. Thus we pass from the subtropical garden to a garden of some other kind, in which "beauty of form" is the prime consideration, and the money value, equally with the native habitat of a plant, are matters of no consequence at all. But, then, if we get away from the ori- ginal idea, we may as well abandon the ori- ginal designation. The proper name for the style of gardening which employs tender and hardy plants alike, and cares nothing for the country, kindred, or money value of any while sensitively an- xious for beauty of a distinctive kind—the proper name for such a style is the Pictu- resque Flower Garden; for picturesque effects are aimed at, and flow- ers are always required to light them up. Let us

SORGHUM BICOLOR.

henceforth call this the Picturesque Flower Garden, as dis- tinct from the highly-coloured flower garden, and the larger subject of decorative gardening which embraces all styles and

15

fashions, will be advanced a stage at least in its nomen-
clature. This is a subarctic climate certainly, but one of the
most wonderful of its class on the face of the earth, if we
may judge it by the thermometer on the one hand, and by
the wonderful and vast variety of vegetable beauty which it
is found possible to display in our gardens, brief as may be
the period during which many of them continue in a pre-
sentable and effective condition.

MUSA CAVENDISHI.

In the construction of a subtropical garden it is desirable
to select a sheltered spot lying open to the south, and to
separate it by suitable planting from other parts of the
grounds to constitute a separate feature. In design it may be
either extremely simple or highly fantastic, provided that the
plants are so placed that they will be likely to thrive and so be
worth seeing, and also that the design is favourable to the dis-

play of their several characteristics. As the beds in which the more tender subjects are disposed will be raised above the general level, and screened by belts of trees and shrubs, a few bold open slopes of grass turf moderately enriched with clumps of cannas, erythrinas, ricinus, and ornamental grasses, will contribute materially towards completeness and richness of effect, and afford the spectator breathing space. A certain snugness of arrangement is as essential to enjoyment of the display as to the prosperity of the plants. The subtropical garden should

be a "shut-in" place, with interrupted entrances to prevent any rush of cold currents of air, opening inward, upon grassy dells sheltered with lime, beech, plane, and a few of the most elegant conifers, as in the sketch subjoined. The principal display of subtropical plants should be on belts and banks next the boundaries, and the more highly-coloured masses on the outer portions of the lawns. The artist who will have the courage to introduce groups of hollyhocks, with belts of cannas associated with them to hide their legs, in the

central spaces, will be pretty sure of enthusiastic plaudits. At
Cashiobury, the subtropical garden is a simple circular space
enclosed by a sloping bank of shrubs, above which is a belt of
mixed deciduous trees. The principal display consists of three
small circular beds (1 and 2, 2) standing two feet above the
ground level, filled with cannas, caladiums, solanums, and

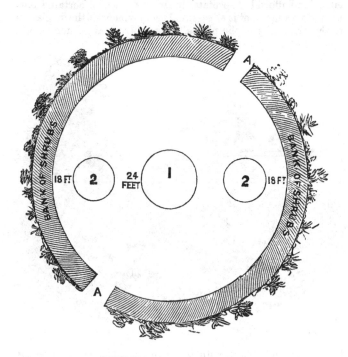

DESIGN FOR SUBTROPICAL GARDEN.

other plants of striking character. The remainder of the space
is grass turf dotted with clumps of pæonies, delphiniums, and
other showy hardy plants.

As we admit to the picturesque garden any suitable plant,
no matter whether hardy or tender, it follows that general
rules for the management of the various occupants of the

DAHLIA IMPERIALIS.

garden cannot be given. We may, however, very safely offer some general advice. As this work is addressed chiefly to the

proprietors of gardens of limited extent, we cannot err in advising the adoption of a few of the most strikingly handsome
and most easily managed plants for subtropical and picturesque
effects, to the exclusion of a myriad of so-called " subtropical
plants " that have been employed in Battersea Park and elsewhere, under circumstances far more favourable to success
than we dare to hope for in the average of well-kept private
gardens. In a comprehensive review of suitable subjects, we
naturally divide them into two classes—those that must be
wintered under glass, and those that are strictly hardy, or so
nearly so that a little rough protection suffices to carry them
through the winter safely.

PICTURESQUE PLANTS OF TENDER CONSTITUTION.—The best of
these for the decoration of a small garden are *acacias, agaves,
cycads, solanums, wigandias, caladiums, palms, dracœnas, musas,
araucarias,* all of which are of the greatest value for the adornment of the conservatory during winter. Those of soft quick
growth, such as solanums, wigandias, and caladiums, would
do best planted out, but the slow-growing plants of hard texture would, as a rule, be more safe if plunged in pots.

It often happens that where stove and greenhouse plants
are grown, a certain quantity of surplus stock burdens the
hands of the cultivator from time to time, and it is a convenient way to get rid of it, without any shock to the feelings,
by planting it out and leaving it to perish. It would of
course be better to throw the plants on the rubbish-heap in
the first instance, if they do not happen to be suitable for the
embellishment of the garden, but in the case of a few surplus musas, begonias, and dracœnas, for example, which may
occasion a little perplexity, the difficulty is disposed of by
planting them out about the end of June, to make a few
novel and dashing effects in the flower garden until the chill
of autumn disposes of them, and saves their owner the pain
of stamping them under foot. In many places where sub-tropical gardening is not systematically followed, occasional essays
in that direction may be made, and irrespective of surplus
plants one may wish to get rid of, many of the most valued
inhabitants of the conservatory and greenhouse may be bedded
out on the plunging system to enrich the lawn, and be vastly
benefited by exposure in the open air during the most favourable season of the year. Viewed from this point merely, there
is simply no end to the possible selection of plants for the

purpose, for even the humblest occupants of the stove and
greenhouse may be turned to account, to afford relief and
support to more stately subjects.   We must be careful, however, to repeat that it
is not tenderness of
constitution that ren-
ders a plant suitable
for the picturesque
garden ; it must have
character of some
sort, if even it be
only grotesque, and
it must be capable of
withstanding a small
gale, a smart show-
er, and a few chilly
nights, without col-
lapsing into rags and
litter.

PICTURESQUE
PLANTS THAT ARE
NEARLY OR QUITE
HARDY.—The range
of selection amongst
plants of this class
is immense, for we
have but to look for
distinctive forms of
leaf beauty, or of leaf
and flower combined,
in plants notable for
elegance or massive-
ness, and sufficiently
hardy to need at the
utmost nothing more
than frame protec-
tion in winter. The
variegated *Aspidistra*

ASPIDISTRA LURIDA.

may be regarded as a good type of the class, a noble, tropical-
looking plant, that endures without harm our severest
winters, but attains to fullest development only when aided
with protection in winter and a kindly heat in spring, when

its new and handsome leaves are in process of development. Several fine *Bambusas* will endure our ordinary
winters in the southern counties, and need only the shelter of

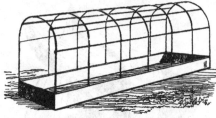

a cool house in the
midlands and the
north. As they are
extremely elegant,
it is advisable, if
there be any doubt
of their hardiness,
to pot them before
they suffer by frost,
and consign them
to the conservatory

PROTECTING TENDER PLANTS IN SPRING.

for the winter. *Cannas* are among the most handsome plants
available for grand effects, and they offer immense variety of
leafage and flower, some
of them gigantic with
leaves of the most deli-
cate pale green, or deep-
est purple, or blackish
bronze; others dwarf in
growth, and equally
various in decorative
characters. Probably all
the cannas in cultivation
may be preserved through
the winter in the open
ground with the aid of a
protecting coat of litter;
and it is certain that a
large proportion of the
most handsome may be
treated as nearly hardy,
for they have been left
out seven or eight years
continuously at Battersea
Park, and a considerable
number are found to be
quite hardy in Paris.

THAPSIA DECIPIENS.

They need a rich deep soil, with abundance of water all the
summer. For their protection in winter a cradle covered

with mats, as in the subjoined figure, will be found more
effectual than litter. *Ferulas* are good enough for choice
positions, but the gigantic umbelliferous plants usually recommended for the dressed garden, such as *Thapsia decipiens* and
Heracleum giganteum, are too coarse, though distinctive and
noble when growing in half-wild places. The *Pampas grass*,
and the larger kinds of *Arundo, Elymus, Andropogon, Holcus,
Sorghum, Zea*, and some other showy grasses, may be employed
with admirable effect, the annual kinds needing to be raised
from seed sown in mild heat in March, in order to insure a
free growth by the time they are planted out. *Gunneras*
require a deep, rich, moist soil, and are quite hardy, a group
of them on the margin of a pool, with tufts of *sugar-cane* and
papyrus rising between, would constitute a striking feature at
the bottom of a dell, or on the margin of a stream. The
gigantic and remarkably elegant *Dahlia imperialis* requires a
sheltered spot in which to make its summer growth, and
should be taken up about the middle of September, and be
carefully potted and placed in the warmest corner of the
conservatory, where it will produce its magnolia-like flowers
in the month of November.

The *Phytolaccas* are quite hardy, and though neither noble
nor elegant, will be found useful to form dense green back-
grounds. *Polygonum cuspidatum* (*P. Sieboldi*) is as hardy as
any British weed, and one of the most distinct of picturesque
plants, particularly well adapted to stand alone on the turf.
Tritomas need not be eulogized, for they are sure to find their
way into the front line in the herbaceous garden. *Yuccas* are
particularly well adapted for our purpose, and they comprise a
few, such as *Y. pendula* and *Y. gloriosa*, that are indispensable,
both because of their striking characters and perfect hardiness.
Others, such as *Y. aloifolia*, both green and variegated, are no
less valuable to adorn the conservatory in winter than to
occupy commanding positions in the garden during summer.

Two capital illustrations of the wealth of material among
hardy plants adapted for picturesque effects were offered in a
picturesque garden last summer. In two far distant parts of
the garden isolated plants of *Crambe maritima* (common sea-
kale), and *Crambe cordifolia* (the heart-leaved sea-kale), were
advantageously placed for full development, and the display
of their very distinct characters. Each plant covered a space
about two yards across with gigantic glaucous leafage, which

in the month of June was covered with a magnificent sheet
of white flowers. Men who had for fifty years grown sea-
kale for a living, scarcely recognized their old familiar friend
of the kitchen garden when promoted to stand alone as an
ornamental plant on a piece of fine turf, surrounded with
valuable and beautiful plants from many climes.

A SELECTION OF SUBTROPICAL PLANTS.

SEVENTY ORNAMENTAL-LEAVED PLANTS, RANGING FROM TWO
TO TWELVE FEET IN HEIGHT.—*Acacia lophantha, Agave Americana variegata, Alsophila australis, Andropogon giganteum,
Aralia papyrifera, A. Sieboldi variegata, Arundo donax versicolor,
Bambusa arundinacea, B. himalaica, Bocconia cordata rotundifolia, Caladium atrovirens, C. esculentum, C. Javanicum,
Canna Annei rubra, C. Auguste Ferrier, C. Barillei, C. Biho-
relli, C. expansa, C. musa hybrida, C. Maréchal Vaillant, C.
rubra superbissima, Centaurea gymnocarpa, Chamœpuce cassa-
bonœ, Cupania filicifolia, Cyathea arborea, C. dealbata, Cyperus
papyrus, Dicksonia antarctica, Dracœna australis, D. cannœ-
folia, D. Cooperi, D. draco, D. ferrea, D. indivisa, D. terminalis,
D. cylindrica, Ferdinandia eminens, Ficus elasticus, F. Por-
teanus, Fourcroya gigantea, Lomatia ferruginea, Melianthus
major, Monstera deliciosa, Musa ensete, M. Cavendishi, Nicotiana Wigandioides, Phormium tenax, P. tenax variegata,
Polymnia grandis, Rhopala australis, R. corcovadensis, R.
glaucophylla, Ricinus albidus magnificus, R. Obermanni, R.
sanguineus, Saccharum officinarum, Solanum acanthocarpum,
S. giganteum, S. marginatum, S. Warcewiczoides, Sonchus
pinnatus, Theophrasta imperialis, Wigandia caracasana, Yucca
aloifolia variegata, Y. filamentosa variegata, Y. quadricolor,
Zea caragua, Z. Japonica.*

TWENTY-FIVE HARDY ORNAMENTAL-LEAVED PLANTS.—*Acanthus latifolius, A. mollis, Aralia Sieboldi, Arundinaria falcata,
Arundo conspicua, A. donax, Aspidistra lurida variegata, A.
elatior punctata, Bambusa metake, Bocconia Japonica, Canna
limbata, Carduus tauricum, Cineraria platanifolia, Crambe
cordifolia, Elymus arenarius, Ferula gigantea, Gynerium
argenteum, Helianthus macrophyllus giganteus, Heracleum giganteum, H. sibericum, Laportea crenulata, Melanoselinum decipiens,
Rheum Emodi, Rudbeckia Newmanni, Salvia chionantha, Yucca
glaucescens, Y. gloriosa, Y. recurva, Zea caragua, Z. Mays.*

TWENTY-FIVE FLOWERING PLANTS ADAPTED FOR SUBTROPICAL GARDENING.—*Abutilon striatum venosum, Brugmansia Knighti, B. sauveolens, Canna picturata fastuosa, C. Rendatleri, Datura*

ECHEVERIA SECUNDA GLAUCA.

arborea, D. fastuosa flore pleno alba, D. Huberiana, Daubentonia magnifica, Erythrina crista-galli, E. laurifolia, E. Marie Belanger, E. ruberrima, Helianthus argyrophyllus, Hibiscus ferox, H. rosea sinensis, Lupinus arboreus, L. mutabilis versicolor, Malva Californicus, Nicotiana macrophylla gigantea, Pent-

DRACÆNA CYLINDRICA.

stemon barbatus Torreyi, Senecio Ghiesbreghti, Solanum verbas-cifolium, Tritoma grandis, T. uvaria glaucescens.

TWENTY-FIVE PALMS ADAPTED FOR SUBTROPICAL GARDENING. —*Areca Baueri, A. monystachys, A. sapida, Brahea calcarea, B. dulcis, Chamœdorea Ernesti-Augusti, Chamœrops excelsa, C. Fortunei, C. humilis, C. palmetta, Cocos australis, C. campestris, C. Wallisi, Corypha australis, Diplothemia maritimum, Latania borbonica, L. rubra, Molinia chilensis, Phœnix dactylifera, P. reclinata, Sabal Adamsoni, Seaforthia robusta, Thrinax parvi-flora, T. tunicata, Trithrinax Mauritœformis.*

TWELVE SUCCULENT PLANTS FOR CARPET AND EMBROIDERY BEDDING.—*Echeveria metallica, E. secunda, E. secunda glauca, E. sanguinea, Sedum Sieboldi variegatum, S. Japonicum varie-gatum, S. azoideum variegatum, S. spectabilis, Sempervivum Californicum, S. tectorum* (common houseleek), *Rochea fal-cata, Mesembryanthemum cœsium.*

- SUBTROPICAL PLANTS THAT MAY BE RAISED FROM SEED.— *Acacia lophantha, A. Julibrissin, Amaranthus melancholicus, A. tricolor, Arundo conspicua, Bocconia cordata, Canna* in variety, *Beta cicla, B. Braziliense, Cineraria platanifolia, Chamœpuce diacantha, Crambe cordifolia, Ferula* in variety, *Ferdinandia eminens, Helianthus orgyalis, Heracleum gigan-teum, Humea elegans, Nicotiana* in variety, *Phytolacca decan-dra, Ricinus* in variety, *Solanum* in variety, *Zea* in variety.

The subjoined figure represents a cheap frame adapted for "hardening" subtropical plants of large growth, such as cannas, palms, musas, etc. It is formed of light woodwork, covered with canvas, with tarpaulin roof wholly removable.

CHAPTER XIII.

THE PERPETUAL FLOWER GARDEN.

There has been enough said in these pages upon the short-comings of the prevailing system of embellishing gardens, and we may turn from the negative to the positive, in hope of some advantage to our readers. We propose, then, to unfold to them a plan for the perfect abolition of tameness and sameness, for making an end of monotony and wearisomeness, for the termination of the floral see-saw, the feast and fast system, by which we make sure of flowers during June, July, and August, and of a beggarly account of empty beds during the remaining months of the year. We are to propound the Arcanum—the secret, the mystery—which is to be no mystery by the time we have done with it; and it is all to be made so plain and pleasant, that from this time forth garden grumblers are to cease from off the earth, disappointments are to be known no more, and the reign of concord and flowery bliss is to set in with such severity as to overcome all obstacles. You are now expecting something new, yet Solomon has averred that there is nothing new under the sun. So beware!

The arcanum to be expounded is the PLUNGING SYSTEM. It cannot be our invention, because plunging in some sort of way was done before we were born. But we claim to have discovered and developed the full possibilities of the system, and profess to know more about it than any practitioners of gardening in all the world. The object of the plunging system is to keep up a rich display of flowers or leaves on the same spot the whole year round, and this is accomplished by growing suitable plants in pots, and plunging them where required when they are at their very best.

The plunging system is nothing unless there are at least four changes in the year—say in April to put out Hyacinths and Tulips, and in May or June to put out Geraniums, Cal-

ceolarias, and Mixtures; in October for Chrysanthemums, and
in December for Evergreens. But there may be twelve,
twenty-four, or even fifty-two changes, if it is the taste of the
proprietor to encourage change, and he has the means of
keeping the wheel turning at that rate. What one may do on
a small scale another may do on a large scale; and wherever
the plunging system is fairly tried, it will be found to surpass
in splendour, certainty, and variety, every other system that
can be thought of to compete with it.

A WINTER GROUP ON THE PLUNGING SYSTEM.

Let us endeavour to give an idea of the system as practised at Stoke Newington. There is a centre circular bed
enclosed in a beautiful jardinet of Ransome's imperishable
stone, and there are three borders, all of them faced with a
handsome moulded curb, also in Ransome's stone. Two of
the borders are planted with trees and shrubs, the principal
border of the three being as richly furnished as possible with
Aucubas, Hollies, Yews, Berberis, Box, Japan Privet, and

other first-class evergreens. During winter this plantation is still further enriched by plunging amongst the permanent shrubs pot-plants of Cupressus Lawsoniana, pyramid Ivies, Irish Yews, and other characteristic plants, all of which are removed in spring to better quarters to promote their growth for the season, as the scene of the plunging is very much overshadowed by large trees. The front lines of these borders, and the circular stone bed, consist of cocoanut-fibre refuse two to three feet deep. It is in these front lines that the plunging, *par excellence*, is carried out in the most complete manner, and a display of colour produced at all seasons of the year, the effect of which is greatly heightened by the depth of green and richness of variegated foliage of the background.

Two remarks are proper at this point. In the first place, well-grown pot-plants, plunged in cocoanut-fibre, have a much brighter, a much more artistic and finished appearance, than plants of the same kinds equally well grown in the open ground. The beautiful, clear, reddish-brown colour of the fibre refuse, by contrast, brings out every tint of green with peculiar brightness, and affords relief to every kind of flower. There is a peculiar charm about a well-furnished plunge bed if the material consists of cocoanut-fibre or clean tan; it is owing to the colour of the material, which sets off and brightens every scrap of vegetation, to which it serves as a groundwork. An amateur who has a passion for floriculture, and is compelled to reside near a town, and must put up with a small garden, may have full gratification of his taste by following the plunging system, and may soon have better collections of plants than the majority of people possessing large gardens, and making pretensions to large practice. Moreover, the system is admirably adapted to produce splendid effects by means of the cheapest plants, and a very large proportion of the subjects grown ought to be hardy, and adapted to bear some amount of rough treatment.

Now let us suppose some one of our readers anxious to carry into effect these proposals; with him or her the question will probably be, "How am I to begin?" We will endeavour to answer the question in such a way as to suit a majority of cases. The first thing to be done is to select the site for the operations, and here a word of advice may be useful to this effect—feel your way carefully, begin with one border or so,

and extend the system as you become accustomed to it, and
equal to its demands, for it will swallow up many more plants
than you have been accustomed to provide for the same space
when planting out was followed.

If we had to advise in particular cases, we should frequently
turf over many of the existing flower-beds, and reduce the
area for display to very
circumscribed limits ; for
in many small gardens the
multiplicity of flower-beds
is puerile, and makes one
think of a doll's garden,
or a farthing kaleidoscope.
Of course we get into
difficulties at this point ;
people are not prepared
to give up their flower-
beds, and do not quite
see the way clearly to do
anything with them but
as they have been accus-
tomed to do.   If there
are groups of beds, and
the desire is to improve
the garden and reduce the
extent of bedding, and
make a first start in
plunging, it will probably
not be difficult to mark
off certain of the beds to
be planted with evergreen
and flowering shrubs,

SEDUM SPECTABILE.

with some good hardy
herbaceous plants in front of them, and reserve the remainder
for experiments in plunging. Let us illustrate this suggestion
by a rough and ready example. Suppose a group of beds, as
in the annexed diagram. We have here ten beds, and we
desire to reduce their number without making them one-sided.
We have but to strike out, say, 2, 4, 7, 9, and we have six
remaining.

Or we may strike out 5 and 6, or 1, 2, 3, 8, 9, 10.

Now, suppose that we cannot attempt to manage six beds by

16

plunging, as shown in the second diagram, why not plant
5 and 6 with groups of hollies, or, if equally convenient (as it

	1	2	3	
4	5	6		7
	8	9	10	

may be in a peat district), with hardy rhododendrons and
azaleas, or with pampas grasses and tritomas, and a few
other such striking and graceful plants, reserving the four

	1		3
	5	6	
	8		10

outside beds for the flowers. This diagram does not illustrate
anybody's garden, but is intended to explain how easily the
way to reform may be found by those who have reforming
tendencies.

Plunging in common earth, that is to say, in the soil of
the place, is possible, but not desirable. So we may use saw-
dust, or old tan, or even moss, or coal ashes. But there is
nothing half so good as the cocoa-nut-fibre refuse; it is always
clean and moist, never wet, never dry, pleasing to look at (as
before remarked upon), harbours no vermin, and a lady
careful of her hands may work at plunging pots in it, and
scarcely find one stain upon her fingers when the work is
done. The next best thing is tan; the next best, moss.
Plunging in mould is allowable, but not advisable; but coal
ashes are simply filthy, and to adopt them in the "plunging
system," that is, as an element in a decorative system, is
heresy. With cocoa-nut and tan there is no need at all to
make provision for the drainage of the pots, but in plunging
in common mould or coal ashes, it is necessary to place a
brick, or an empty inverted pot, under every pot containing
a plant, to prevent the plant becoming water-logged, and also
to keep out worms.

The question now is about the formation of the plunge
beds. In places where stone or wooden edgings are already

in use, there is not much difficulty. You decide what is to be
the width of the plunge border, and to that width the earth
is to be dug out. If the border is narrow (say three feet), a
depth of eighteen inches will be enough, because very large
pots will not be used. But if wide (say six feet), it may be
cut to a sloping bottom twelve inches deep at the extreme
front to three feet deep at the extreme rear, which will allow
of the largest pots or tubs with specimen conifers for the
back row in winter time. In some places good plunging will
be done with small pots, and in other places good plunging
will be done with large pots; and again some practitioners will
indulge largely in winter trees, and some will only care for
summer flowers, etc., etc. Where beds are cut in grass, it is
an easy matter to take out the earth and put in suitable
plunging material; where there is a grass verge to a border
there can be no difficulty in cutting sharp to it; but in case
of a box or thrift edging, the cutting must be done with care,
or the edging may be killed. Put down the line three inches
from the live edging, and cut down sloping, so as to spare the
roots. If flooring boards, or any rough planking, can be
afforded, line the bed with timber, back and front, as shown
in the diagram, where we suppose the front to be clipped

Edging.	Plank.	Plunge Bed.	Plank.	Soil Planted.

box, or in any case a bold and substantial stone edging; next
within that, as a lining, a plank on edge; then a given
breadth of cocoa-nut-fibre refuse for plunging; next a plank on
edge as before, and then, beyond that, the undisturbed soil of
the garden, with a background of evergreens, etc., etc.

When all this is done, there must be established a regular
system of cultivation to keep the beds supplied. If this
cannot be accomplished, better no plunging at all. However,
one or two borders may be tried at first, and the system of
growing will be found to be more simple than appears; and,
in fact, its chief characteristic is that *it is a system;* every
separate batch of plants must be prepared to come on in its

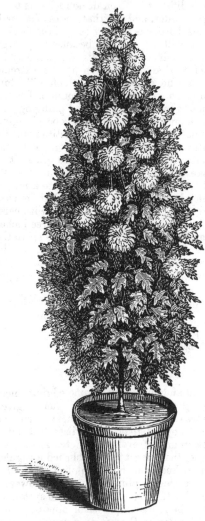

PYRAMID CHRYSANTHEMUM.

proper time, with no excessive glut to bewilder the cultivator, and never a deficiency of good things to make a cheerful display on any day in the whole round of the year.

The principal subjects for plunging are, for early spring, aconites, snowdrops, crocuses, hyacinths, and tulips: for late spring, wallflowers, yellow alyssum, white iberis, rosy aubrietia, sparkling dielytra, bold and handsome crown imperials. For early summer, stocks, roses (brought on in pits or by slow forcing), yellow cytisus, deutzias flowered in cold pits, rhododendrons, and a few of the more showy annuals grown in frames. For succession, geraniums, calceolarias, and all the rest of the summer flowers. For September, Sedum fabarium; for October, British ferns, then all fresh and bright, with any odds and ends of colour to light them up. For

November, pyramid and bush chrysanthemums ; for December,
ivies, conifers, and berry-bearing shrubs, and so on to the
spring bulbs again.   In selecting subjects, and in the cultivation of the plants, it must be always remembered that
spreading concave-headed plants are of less value than comparatively narrow, and in the case of zonal geraniums, " long-
legged " plants, because of the rather close packing required
to produce a rich effect.   Thus the pyramid chrysanthemum
figured on page 244, is far to be preferred to the dwarf, close-
-trained, convex plant that would suit the parlour window.
The tall, spare habit, and fresh appearance of well-grown
seedling geraniums render them invaluable for plunging.

CHAPTER XIV.

THE ROCKERY AND ALPINE GARDEN.

THE course of procedure sanctioned by custom in the literary treatment of this subject consists in first destroying all existing rockeries everywhere by unqualified abuse, and then reconstructing them on the author's model, on the hypothesis that they do not exist to please their owners, but to illustrate the writer's theory of what a rockery should be. We beg permission to evade the demands of "tyrant custom," and avoiding controversy, to find opportunity here for a few suggestions that may be useful to the reader.

There are many kinds of rockeries, and they serve many purposes. They are sometimes intended as mere screens to hide from view objectionable objects ; but usually they are adopted for purposes of ornament, and to afford their possessor suitable situations for the cultivation of ferns and alpine plants. As a matter of taste, a really "savage," or say rustic rockery, should not be associated with straight walks, smooth lawns, vases, statuary, and clipped trees. Yet in the most finished part of a garden, a modified form of rockery may be admissible, as, for example, a circle of large unhewn stones to form the boundary of a fountain, with calandrinias, portuluccas, mesembryanthemums, and variegated-leaved plants of trailing habit dotted about amongst them. But this would not be a rockery, properly speaking ; it would be a garnishing of an artistic scene, with rocks only partially displayed, but affording a suitable groundwork for the flowers displayed above them. A rustic rockery may be made a most interesting feature, and a good connecting link between distinct scenes, while at the same time it provides sunny, shady, dry, moist, and marshy sites for an interesting assemblage of beautiful plants. When associated with a resting-place and reading-room, affording shade and coolness in the heat of summer, and with open sheltered seats facing south for enjoyment of the landscape, and the song of the lark in the early spring, and in all seasons providing entertainment in its varied vesture of ferns, mosses, flowering plants, and picturesque surroundings, a rockery may indeed become a most delightful accessory, acceptable alike to the botanist, the artist, and the eclectic idler, whose desire it may be, having escaped from a world of cares, to "rest and be thankful." If far enough removed from the dwelling-house to be consistent, a cottage, or miniature chatelet, comprising a reading-room, a smoking-room, and a chamber suitable for accommodating luncheon and tea parties would be a most valuable aid to the enjoyment of a good rockery. A model of a ruin on a sufficiently large scale could be equally well adapted for social gatherings and meditative retirement, and in a place of sufficient extent to admit of it without breach of propriety, the ruin and the hermit's cell, if adopted at all, should be thoroughly developed and made a delightful place of retreat.

ERODIUM MACRADENIUM.

The materials of which rockeries are constructed must vary in different localities. If the district affords suitable stone, and large blocks can be placed to represent natural protrusions of the strata from below, there will be a great gain in reality and force, provided these strong features are not weakened by puerile accompaniments, such as paltry piles of shells, and cairns formed of the sweepings of a builder's yard. It is the very first requirement of taste, no less than of the most sober common-sense, that the natural capabilities of a place should be developed, and in a rocky country the materials most readily available are almost certain to be the most suitable, because in harmony with the *genius loci*, while such a tempting site as an old chalk quarry, or disused gravel pit, may be turned to grand account for the cultivation and display of ferns and alpine plants in picturesque arrangements. In the neighbourhood of great towns, and especially about London, the best available material is the product of the brick kiln, and what are there called "burrs" answer admirably, for they may be built into any form, and when the roots of plants come into contact with them, it is to the advantage of the plants, rather than otherwise, which cannot be said of the glassy and impervious furnace slag, and other vitreous substances that are occasionally employed.

In constructing a rockery or ruin, definite measures must be adopted to provide sites for plants. Mere handfuls of soil on the tops of dry walls may suit a few of the hardy succulents, but it is of the first and last importance that there should be large masses of suitable soil in all parts of

the structure which it is intended to embellish with plants, and especially for hardy ferns and alpines. Almost any plant will live for a season or so in a spoonful of mould, if watered twice a day, and watched like a criminal at large; but if plants are to thrive in a rockery, they must be encouraged to strike their roots deep into a soil adapted to their nature, and there should be no stint of stuff to promote deep rooting when the work is in process of construction. A very large number of fine rockery plants will thrive in the most common soil with an infinitesimal amount of attention, and in the most off-hand way we may treat such subjects as arabis, alyssum, campanula, asperula, cerastium, corydalis, iberis, and a hundred others. But for the best of the hardy ferns there must be an ample bed, or many beds, and masses of sandy peat; for a few of the rock-loving ferns, shelves and

clefts filled with sandy peat and broken freestone; and for the
majority of true alpines, such as androsaces, the smaller dianthus, erinus, erythræa, hepatica, iris, lychnis, myosotis, and
others of like nature, sandy loam of a mellow texture, and
rich in fibre must be provided, and much of it hidden away
beneath the masses of rock in such a manner that the roots of
the plants will find it by fair searching in a perpendicular or
oblique direction downwards—it must not be expected that
they will turn corners and go upwards after it anywhere. A
genuine rockery in the garden of a genuine amateur should,
in the first place, have one distinct character of its own, and
an evident set of rela-
tionships. For example,
if it were determined to
construct a ruin, it should
be the ruin of a sup-
positious castle, church,
keep, or part of one, and
should be constructed of
one material throughout,
or at least not of an in-
congruous mixture of
materials, such as the
original   suppositious
builder would not have
employed.   If, on the
other hand, a natural
pile of rocks were imi-
tated, it should consist
of such rocks as might
be met with somewhere,

THYMUS AZUREUS.

and not of a collection of geological curiosities in the fashion
of an outdoor museum. The burrs from the brick-kiln answer
admirably for a bastion or keep, or for an irregular construction of miniature mountains, valleys, and gorges, but it would
not be in good taste to dot the slopes with shells or stick busts
of eminent (or unknown) personages on the pinnacles. If of
sufficient extent, it should present a number of aspects and
considerable variety of conditions, such as rough terraces and
knolls facing the north, for the smaller and more delicate
alpines, which thrive the more surely when enjoying a cool
climate all the summer, and are actually safer in winter in

the coldest, rather than in the warmest place, not because
long-continued frost is any particular benefit to them, but
because the bursts of bright weather we usually have in early
spring tend to hurry them into growth too soon, and they may
afterwards suffer through the return of frost, snow, miserable
rain, or keen, parching east winds. It is just because of the
variableness of our winters that alpine plants are frequently
grown in alpine houses, which are low-roofed, brick-built
greenhouses without heating apparatus of any kind, affording
shelter only, and saving the delicate alpine plants from the
destructive influence of intense cold following unseasonable
warmth, and long-continued rains, accompanied with forcing
weather when they should be still and quiet under a covering
of snow.

But all this far-reaching scheme of a rockery may suit but
few of our readers ; nevertheless, the principles are the same
in the construction of a small as of a large rockery, and for
just a moment we will peep at one of the smallest, which
happens also to be one of the best with which we are familiar
in our daily walks. This consists of a mere bank of common
loam heaped up against a cottage wall, and faced with bricks
and burrs that the cottager gathered from the road-side bit by
bit, and saved until he had enough. It is crowded with beautiful plants, and is in all seasons a most elegant adornment of
a rustic artist's home. In the depth of winter it is fringed
with the golden-tipped stonecrop, which then, owing to the
yellow colour of the tips of the shoots, is nearly as gay as
the common stonecrop is when in full flower in the height of
summer. As the spring advances, tufts of saxifrages, primulas, drabas, cerastiums, arabis, and alpine phlox burst into
flower, and make a brilliant enamelling of snow-white stars,
rosy cups, blotches of gold and silver, and tiny sheets of
purple and pink. These are followed by bonny clumps of
campanulas, arenarias, tunicas, armerias, aubrietias, corydalis,
linarias, the alpine lychnis, the hoary blue-flowered thyme,
and tufts of the lurid red flowers of the common houseleek and
the very delicate and pretty spider's-web houseleek. It is not
wanting in flowers in autumn, but it is more enjoyable then
for its sumptuous garniture of moss-like verdure of many
delicate shades of green which the saxifrages contribute,
varied with patches of grey and golden leafage. No one
could determine from its appearance that it consists of

veritable rubbish below, because it is so richly clothed, that the bricks and stones employed in the building are hidden, and it completely accomplishes its purpose, that of affording a suitable site for a number of beautiful plants that require no costly aids to their development, but are better adapted for a raised bank of soil and a stony surface to rest upon, than for the common border, which is usually too damp in winter for herbaceous plants of a trailing, surface-spreading, or cushion-forming character.

In selecting rockery plants, it will be well for the beginner to avoid all expensive and troublesome subjects. There are numbers of plants of noble character that are not usually classed as rock plants, that may be introduced advantageously for distinctive effect in bays, recesses, and commanding heights, such as the pampas grass, the arundo, bambusas in variety for sheltered nooks, the gigantic heracleum, the acanthus, the lyme grass, and the phytolaccas. Then for rich colouring, numbers of cheap and showy herbaceous and sub-shrubby plants, especially such as spread laterally, as for example, the dwarf hypericums, the double yellow lotus, the double dyer's broom, the alyssums, iberis, the hardy geraniums, erodiums, campanulas, saxifrages, se-

HYPERICUM PATULUM.

dums, and every species and variety of thyme that can be got. Hardy variegated-leaved plants are especially valuable, and trailing plants, such as the ivy-leaved toad-flax, the periwinkles, ivies, and the golden-leaved and common moneywort are indispensable. For the shady places there are hardy ferns and equisetums in endless variety, and the lily of the valley is not the least worthy to associate with them on open slopes, where the sun peeps in morning and evening to diffuse a genial warmth, favourable, for the most part, even to shade-loving plants.

As for the alpines proper, the cyclamens, androsaces, mountain pinks, droseras, epimediums, and gentians—to name a few only as examples—the best advice that can be offered to the amateur is to acquire experience in their management patiently, without haste or any costly experiments, for a serious disappointment at the first start may seriously damp the ardour under which the start was made, and make an impression unfavourable to the pursuits of the higher departments of decorative gardening. It may be said of alpine plants in general, that to plant them properly in the first instance is everything, and to manage them afterwards is nothing, for, as a rule, if they are but kept free from weeds, and left alone, they will acquire a firm hold of their positions, and do full justice to their owner's taste in selecting them.

The association of water with a rockery is eminently desirable if it can be accomplished conveniently. Were it possible to occupy much, instead of little, space with this subject, something might be said of a stream that passes through a certain garden, and, being dammed up where a slight fall occurs, forms a miniature lake, on which a pretty lot of water-fowl disport for the embellishment of the scene, and for the occasional embellishment also of the dinner-table. Beside the rustic bridge across the dam rises a rustic tower richly clad with ivy, and portions of a ruined wall and broken arches representing a ruin. In all the nooks, and on the walls, and everywhere about the spot, alpine plants, mosses, ivies, and snapdragons run riot, and charm the eye, whichever way it turns. Concealed in the top of the tower is a great tank, and under the bridge is fixed a water-ram, the duty of which is to pump up water to the top of the tower, and provide the gardener with a constant supply, warmed and softened for use as needed all the summer long. Within the tower is a reading-room, and a cool retreat open to the sky, through a lattice of leaves, lurks amid the arches, and woos the breeze to whisper in passing by, "Here care should be forgot." But we must tell of this rockery some day and somewhere in proper detail, and, as we have dipped the pen in the limpid stream, we will present in brief a scheme from the garden of a friend, which may be useful to many a reader of this volume. It is an ornate rockery, differing very much from such as have chiefly occupied our attention in the fore-

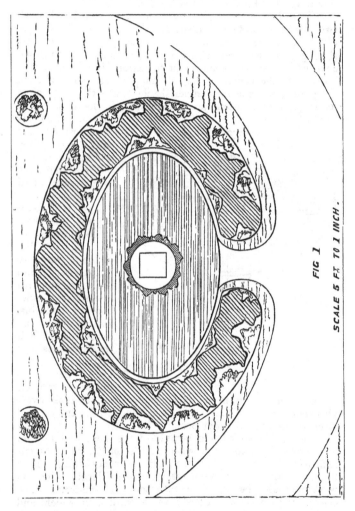

FIG 1

SCALE 5 FT. TO 1 INCH.

going observations, and its place and surroundings may be

discovered by reference to page 9, where it will be found at
figure B in the Plan of a Villa Garden.

We will suppose a piece of ground, twenty-five feet by
twenty feet, to be at your disposal. Mark out the inner oval
(Fig. 1), and excavate three spades deep, leaving the sides
sloping ; then well ram all over, till the surface is firm and
compact. A square pedestal of brickwork in cement is now
built up in the centre to form a support for a vase of iron,
terra-cotta, or Ransome's imperishable stone ; the water-pipe,
for supplying a fountain jet is embedded in the work ; as is
also a leaden ser-
vice, for a range of
minor jets, after-
wards referred to.

Fig. 2 shows the
arrangement of all
the piping.

S. — General
supply of iron,
three-quarters of
an inch bore, fur-
nishing water to
centre vase jet, and
connected at t with

S, s, s, s, s.—Se-
condary fountains,
issuing from an
oval ring of half
inch iron pipe, and
playing into the
tazza. They are re-
gulated by tap (t).

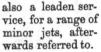

PLAN OF PIPES.

FIG. 2.

5 Feet to 1 Inch.

M, m, m, m, m, m, m, m, are minor jets, receiving over-
plus water from the vase, through half-inch leaden tubing, and
discharging into the lower pool.

The smith's work being completed, the sides and bottom
of the excavation are lined with flat tiles in Portland cement,
and the whole surface rendered with an even coating. That
portion of the brickwork pedestal above the level of the tile-
work must also be plastered with the same material. The
cement is then allowed a few days to harden, in which, and
by way of taking time by the forelock, you had better have

carted to your yard three one-horse loads of " run bricks,"
or " burrs."

The plaster work having set, a narrow rim of turf is laid
round the extreme edge of the pool, thus concealing the
secondary fountain supply.

The nozzles, made of lead, beaten round iron wire, spring
up amid the grass. In arranging the rockwork, commence
and continue in horizontal layers; build up gradually, for-
tifying all weak points with a little gauged cement. The
centre of the back, which is the highest point, should be about
seven feet above ground. Allow plenty of space for good
soil, between the outer and inner walls ; and carry the erection
into jagged peaks, with pinnacles, leaving miniature ravines,
bays, and chinky hollows. Bear in mind that the height must
continue to decrease as you approach the front, where the
greatest allowable altitude will be about eight inches or a foot.
Harts-tongue ferns are planted on a small pile of rockwork
fixed round the square pedestal. The inner nooks are devoted
to ferns. At the edge next the water, the moneywort will
flourish with luxuriance. The outer recesses are filled with
dwarf roses, and the very showiest of dwarf-growing herbaceous plants. Gold and silver fish are placed in the basin,
soon becoming tame enough to flock to the surface for food,
on the approach of their owner, and sporting about among a
few choice water plants, they impart an additional charm to
the already varied scene.

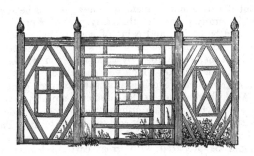

CHAPTER XV.

FLOWERS FOR WINTER BOUQUETS.

The best flowers for winter bouquets are undoubtedly those of the stove and greenhouse, bright with colour, fresh with fragrance, and with the soft and supple texture of active life in them. Genuine winter flowers are a privilege of the few; for the many who cannot obtain them, dried grasses and everlasting flowers are of some service, and may be turned to wonderful account in the preparation of elegant bouquets. We must not waste space in eulogy, but assume that flowers and grasses are required for winter bouquets, and then proceed to the practical business of producing them.

First, as to the cultivation. All the annual sorts, both of everlasting flowers and grasses, are best grown by sowing the seeds in light rich soil in March or April, and treating them just the same way as asters; that is, in brief, insuring strong plants by the middle of May, and then planting them out. But if this is not convenient, they may all be sown on a rich light sunny border, in the early part of April. Every patch should be tallied, and a bit of seed of every sort kept in reserve. About the middle of May sow again any that have not, by that time, come up. By this plan you will be likely to secure all the sorts on which you speculate.

As for greenhouse everlastings, they require good cultivation. As the best of these is the Aphelexis, a practical word on that may be useful. It is a difficult plant to grow, too much or too little water being pretty certain death to it. The soil should be good turfy peat, and plenty of silver sand. The pots should be prepared with great care to insure perfect drainage. The plants must be potted firm, and with the collar slightly above the surface. Plenty of light and air are essential. The beautiful Phœnocoma requires similar treatment. As for the greenhouse Statices, they require a soil half loam and half peat, and a warmer and closer part of the house will suit them than the two plants first mentioned require.

Next, as to gathering the flowers. Take them in all possible stages; but by far the largest proportion should be young and scarcely fully expanded, as they are sure to expand in the process of drying. To dry them, lay them on papers in an airy warm place, *safe from dust*, and store them when dry in dry closets or drawers where dust is as nearly as possible unknown. The grasses may be dried by simply laying them between folds of blotting-paper, or placing them between the pages of large, heavy books. Remember, "practice makes perfect:" the beginner is sure to spoil a few; never mind, there will be many good ones to make amends.

As to mounting, the grasses must be used in their natural state; but it is best to mount the flowers on wires. This is a nice proceeding; but ladies generally acquire the art in haste. The finest steel wire is the best adapted to the purpose, and it is attached to the flower at the base by merely thrusting it into the centre; but the wire should have a few twists to make a sort of base to catch the flower, or for the natural base of the flower to rest on.

The best flowers for this purpose are the following:—

Helichrysums of all kinds; more especially *H. bracteatum, H. compositum, H. macranthum,* and *H. monstrosum.* All are half hardy annuals, to be raised on gentle heat, and planted out in May, or sown in the open ground in April. As they are so useful, it would be well to try all the sorts the seedsmen can supply.

Acroclinium roseum.—Sow in pots and pans in April, and place in cold frame, or sow in open border and risk it.

Rhodanthe Manglesi, R. atrosanguineum, R. maculata, R. major. All half-hardy annuals.

Helipterum Sandfordi and *H. corymbiferum* require careful culture. Sow, if possible, in February, and treat as perilla or lobelia. These are the least likely to succeed if sown in the open border in this country. They are so beautiful that they well repay a little extra care.

Polycolymnia Stuarti.—A quite hardy annual, but none the worse for being pushed forward under glass.

Ammobium alatum is a perennial, but may be treated as an annual, as it is sure to be killed by a sharp frost. Treat it as half-hardy.

Waitzia corymbosa, W. grandiflora, fine half-hardy annuals;

17

but of no use to beginners for winter wreaths. They must be started early to make sure of a good bloom.

Xeranthemum annuum, X. album, X. caryophillioides, and *X. purpurea* are fine hardy annuals, all of which may be sown in the open ground in April. They are not the most desirable, as their colours are apt to fade when dried, but this may be in some part prevented by drying them *quickly in the dark,* and in a very dry, warm atmosphere. Try them in an oven when the fire is nearly out.

BRIZA MAXIMA.

The selection of Grasses may be almost indefinitely extended, and the hedgerows will supply many of the most lovely grasses in the world. The following, however, are worthy of special attention for associating with everlasting flowers.

Stipa pennata is one of the most distinct of all our garden grasses. It grows superbly on a dry, sandy bank, and is adapted for a sunny part of the fernery. What grace, what delicacy, what is there in the vegetable kingdom to equal it for fairy-like elegance? Beware! In the seed catalogues you will see that seed is offered. True, seed *is* offered, but it is comparatively worthless, and the only sure way to secure this grass is to purchase plants.

Agrostis nebulosa, a most elegant grass, having stems as fine as hairs, and fruit panicles so light and "nebulous" that at a little distance a patch of this grass looks like a cloud of vapour. Some seedsmen send out Polypogon Monspeliensis, under the name of Agrostis nebulosa.

Briza maxima is the most useful of the quaking grasses, but all the Brizas are pretty. This grass is much used for winter bouquets, and is invaluable to persons engaged in taxidermy, on account of its suitability for dressing up cases of stuffed birds, etc.

Chloris radiata is a very curious grass, the flowering occurring in a compound spike which consists of five or six separate rays, remotely resembling long fingers.

Lagurus ovatus, a favourite with those who grow grasses for bouquets. It is popularly known as Hare's-tail grass.

Pennisetum longistylum, one of the most elegant grasses known.

PANICUM ITALICUM.

Panicum Italicum is one of the best of a beautiful family. *P. capillare* is also a most graceful species. *P. Miliaceum* (common millet) is also well worth a place in any amateur's garden. Indeed all the Panicums are worth growing. So also is

Setaria Germanica and *Setaria macrochœta*, the last being a thorough " cat's-tail " grass.

Eragrostis elegans cannot be surpassed for elegance when in flower, though until the bloom appears it has a rather coarse appearance.

GROUP OF EVERLASTING FLOWERS.

Milium multiflorum is the most elegant of this elegant family. It is invaluable for winter bouquets to mix with ever-lasting flowers.

Airopsis pulchella, a little gem for pot culture. When covered with seeds it is quite a curiosity.

Hordeum jubatum is the pretty squirrel's-tail grass, a good companion to Lagurus ovatus.

Ægilops cylindrica, a stiff, quaint, and not inelegant grass, which comes in well for bouquets.

Lepturus subulatus, a wiry backbone sort of grass that will make any one laugh who sees it for the first time.

GROUP OF ORNAMENTAL GRASSES.

Bromus brizæformus, a minute grass of the most exquisitely graceful construction. It is a genuine candidate for complete seclusion in fairyland; such a sordid world as this does not deserve to behold its beauty.

The mixed border will supply a few good flowers for drying, such as the hardy statices and gypsophilas, and it is an easy matter to dry the flowers of double geraniums so as to preserve their brilliant colours in perfection.

CHAPTER XVI.

THE MAKING AND THE MANAGEMENT OF THE LAWN.

AMONGST the earliest recommendations in this volume is one in behalf of the greatest possible breadth of well-kept turf consistently with the area enclosed for purposes of pleasure. To insure the luxury of a "velvet lawn," is, to speak generally, a most easy matter, and, though it may be comparatively costly in the first instance, it will prove in the end one of the best of investments of gold in gardening. The soft elastic turf of a chalky down will kindly inform the traveller that a lawn may be laid on chalk; and the closely bitten grassy herbage of a sandy common will in like manner suggest that gravel and sand may be clothed for the production of a living carpet that will last for ever. It is, however, on a deep loam or a clay that has been well tilled, that the best example of grass turf is to be looked for, and on such land we should prefer to operate, were it required of us to present the best possible example of making and keeping a garden lawn.

In the formation of a lawn, all levels must be carefully determined, and the ground thoroughly well prepared, that there may be no waste of labour in alterations afterwards. In the case of laying fresh turf on the site of an exhausted plot, from which bad turf has been removed, a heavy dressing of good manure should be dug in, for grass needs nourishment in common with all other plants. The last act of preparation consists in spreading over the level ground about an inch depth of fine earth, which is to be distributed evenly, and every stone removed by means of the rake. Then we approach an important question—which is best, turf or seeds? In any and every case turf is to be preferred, for upon the instant of its being laid and rolled, the lawn is formed, and there is an end of the matter. Two considerations give interest to this question—the cost of turf is necessarily far in excess of the cost of seeds, and it may happen that turf is not to be obtained within reasonable carting distance. Supposing the amateur to have a choice of means and materials, our advice would be in favour of the purchase of the best turf possible, for any extent of ground under one acre; but when we get beyond an acre, with every increase of extent, the argument in favour of seeds increases in force, for the cutting and carting of turf is a somewhat costly business. In selecting turf for a garden, give the preference to that which is of close texture, containing a fair sprinkling of clover intermixed with the finer grasses. We have formed many lawns from meadow turf, which in the first instance appeared far too coarse, and they have in the course of three years acquired a beautiful texture fit for the foot of a princess in a fairy tale. Grass turf may be laid at any time during favourable weather, but the autumn is to be preferred, because of the long season of growth the newly-laid turf will have to aid in its establishment before being tried by the summer sun. If laid early in the spring, grass usually passes through the first summer safely, but is of necessity exposed to the risk of being roasted; in the event of a hot dry summer, the risk is greater in the case of turf laid late, than of turf laid early. When the work is deferred until the season of spring showers is past, it will be advisable to spread over the turf a coat of good manure, and keep it regularly and liberally watered until showers occur.

In selecting seeds, the character of the soil must be taken into consideration, for a mixture that would suit a clay or loam would not equally well suit a sand, gravel, or chalk soil. The seedsmen who make a "speciality" of grass seeds will for any given case supply a better mixture than any one unskilled in the matter could obtain, even if acting on the advice of a botanist or gardener. As, however, prescriptions are occasionally required by seedsmen who have not had extensive experience, we shall append to this chapter a few for mixtures adapted to particular kinds of soils. The best time in the year to sow seeds is the month of August. If the work cannot then be completed, the sowing may be continued through September and October, but not later; and may be resumed in February and March. Grass seeds may be sown indeed on any day in the year, provided the weather is favourable for the operation, and the ground in a fit state; but the

month of August is the best time to insure a good plant
before winter, and a long period of growth before the summer
heat returns.

There is yet a third mode of forming a lawn, now rarely
practised, but in days when grass seeds were comparatively
unknown, frequently resorted to. It is termed "inoculating,"
and consists in planting pieces of grass turf at regular distances over the plot. In districts where good turf is obtainable only in small quantities, this method may be recommended, for if the turves are torn into small pieces, and
planted at a foot apart in September or February, they will
extend rapidly, and form a pretty good sward the first season.

In the after management, the principal operations consist
of rolling, mowing, and weeding. Grass seeds must be constantly weeded, until the turf thickens sufficiently to kill out
the weeds, and newly-laid turf must be kept clear of thistles,
docks, and other rank weeds, by spudding them out; or by a
simpler process which we have long practised with the most
agreeable results, that of depositing in the heart of the plant
a small quantity of phospho-guano, which kills it at once,
and promotes the growth of clover in its stead. If this
operation is carelessly performed, and the guano thrown
about wastefully, the immediate result is a dotting of the
lawn with unsightly brown patches, which, however, soon
disappear after the occurrence of rainy weather.

Many as are the kinds of mowing machines, they may
all be classed under two heads—those that cut and carry,
and those that cut and scatter. A carrying-machine may
be made to scatter by removing the box, but not so well
as the machine that is intended for scattering, as in each
case the cutter is formed expressly for the work it is intended to perform. If the question be asked, which is the
best form of machine, our reply is that they are of equal
value, and the intending purchaser must be guided by a consideration of circumstances. In the excessively hot and dry
summers of 1868 and 1870, we constantly employed the
"Archimedean," which scatters the grass, and our lawns were
as green through all the burning drought as in the cooler
days of spring. In the moist summer of 1871, it would have
been necessary to sweep up the grass, had the scattering
machine been employed on our strong land, and therefore
we kept our trusty "Shanks" at work, cutting and carrying,

and had to mow twice a-week through the whole of June and July to keep the grass down. Nevertheless, in that same moist summer, we saw the "Archimedean" employed on a tract of chalk land, which is peculiarly exposed to the influence of the sun, and the result was a fresh green turf, where in the height of summer nothing better than a dusty door-mat had ever been seen before. When the grass is cut by cutters adapted for the scattering system, it falls on the ground in a form more resembling dust than fibres, and acts as a "mulch" both to nourish the growth and arrest evaporation from the soil; hence the importance of the scattering system on chalk and sand, and other hungry stuff, and on any soil in such a hot season as that of 1870.

In the keeping of an old lawn it is of the utmost importance to remember that grasses and clovers require for their well-doing a highly-nourishing soil. Now it matters not how good the soil may be in the first instance, if we cut and carry, we labour constantly to impoverish the top-crust. In every barrowful of grass removed, there will be a certain quantity of alkalies, phosphates, and other constituents of vegetation, abstracted from the soil. To be always taking off and putting nothing on, must result in the starvation of the grass; and we shall find that as the grasses and clovers disappear through the exhaustion of the soil, daisies, plantains, knotgrass, self-heal, and other weeds, will take their place. The simple remedy for this state of things is manuring, and the best mode of manuring is to scatter over the turf a succession of thin dressings of guano and fine mould mixed together. This should be done in autumn and spring, at times when there is not much traffic on the grass, and there is a likelihood of rain to follow. If appearances are of no consequence in the later autumn or early spring months, a good coat of half-rotten manure may be spread over the turf, but this proceeding cannot be recommended for general adoption. In place of guano, nitrate of soda or nitrate of potash may be employed, being first mixed with fine earth or sand, and then scattered at the rate of one pound of nitrate to every square yard. The employment of an alkali will promote the growth of grass, but not of clover, which requires the aid of phosphates. A cheap and most serviceable dressing for old lawns may be occasionally obtained in districts where building works are in progress. The rubbish should be screened, to separate from it the dust of old mortar, plaster, and broken brick to the size of walnuts at the utmost. This may be spread thinly two or three times in autumn and spring, and will greatly benefit the texture and density of the turf.

It cannot be said that in British gardens grass is generally well managed and properly understood, for the lawn is the last place on which either manure or water is generously bestowed. We may ofttimes see the flower-beds deluged with water that they do not need, while the grass is fast parching into a hideous condition of sterility. If we could persuade the industrious folks to spread the water, by means of a hose, over the grass two or three times a week during summer, and give the geraniums none at all, the result would be a brighter blaze of flowers in a rich setting of delightfully fresh verdure, instead of, perhaps, geraniums growing like cabbages, and scarcely flowering at all, and the grass becoming as thin and black as if a flame had passed over it.

Two contingencies are to be especially guarded against in the management of grass turf—the machine must be set so as to cut fair, and it must be kept in the best order by constant cleaning and oiling. If set so as to cut very close, it will occasionally pare off the surface soil, and with it the roots of the grasses; many a good lawn has been ruined by the foolish practice of making the machine cut as close as possible, under the absurd impression that one cut is better than two. The more cuts the better, provided always that the machine is properly set and in the best working order. Another mode of making a present effect at the expense of the lawn consists in continually cutting a fresh edge with the edging iron. A gardener who cuts into the turf on the edge of the lawn to make a finish ought to be compelled to eat all that he removes. If the practice is persisted in, the grass is reduced in breadth, and the walk is widened, and in time there is formed a deep gutter and a sharp ugly ridge. If properly finished at the edge with the shears, the width of the walk will not vary an inch in fifty years. One of the first things we look after in the work of a new man is his management of the edges of lawns, and we are always careful to explain our views upon the subject in good time to prevent a mischief which cannot be easily remedied. The man who persists after warning and explanation in chop, chop, chop-

ping at the edge, as if it were necessary to construct a gutter
of mud on each side of a walk, deserves to hear an opinion
of his procedure that will make him tingle from head to foot
with shame. The jobbing gardener is a master of this chop-
down-gutter-forming business, and will always be ready to
advise the employment of gravel to fill up the trench that
should never have been made.

It may be well to add a word upon the employment of
Spergula, or more properly of *Arenaria cæspitosa*, for lawns.
A "spergula lawn" in good condition is one of the loveliest
embellishments of a garden that can be conceived. We have
seen only three that were good enough for agreeable remem-
brance. ·The whole truth of the matter may be summed
up in a sentence : A spergula lawn demands constant attention, and is of necessity a troublesome thing to form in the
first instance, and to manage afterwards. Therefore, for
what may be termed "general usefulness" we cannot recommend the employment of spergula. However, any of our
readers who are inclined to indulge in this unwonted luxury
need not be deterred through supposing there is any mystery
at the bottom of success; it is a question of time and attention, and whenever these are withheld the spergula lawn will
go to ruin. Prepare the ground well, and plant the tufts in
September and October, or in March and April. Frequently
roll the ground, and never cease to pull out weeds, for these
are the chief enemies of spergula. One season's neglect of
weeding will ruin a spergula lawn, and one week's neglect at
a time of year when weeds grow freely will result in considerable damage. As for worms, which occasionally injure
spergula turf by their casts, the roller will sufficiently repair
the damage ; but if any nostrum is required to reduce their
number, there can be nothing better than clear lime-water,
for, while this kills the worms, it benefits the spergula.

As the formation of a spergula lawn requires much patient
attention, it may be recommended as a pastime to those who
are of a temperament suited to the task, and can afford the
time that must be devoted to it for a satisfactory result. Our
advice to a beginner, fired with enthusiasm on the subject,
would be to select a comparatively small piece of ground in
the first instance, in order to obtain a perfect sample of spergula turf in the shortest possible time, and acquire thereby
the experience needful for a greater effort. For those who

practise "rough-and-ready gardening," spergula is of no use at all, except as a rock plant or to cover a knoll.

For all GOOD LOAMY GARDEN SOILS, the best grasses to form a close, fine sward, are the following :—*Cynosurus cristatus*, the crested dog's-tail; *Festuca ovina*, the sheep's fescue; *F. tenuifolia*, the fine-leaved fescue; *Lolium perenne tenue*, fine-leaved rye-grass; *Poa pratensis*, smooth-stalked meadow grass; *Poa sempervirens*, evergreen meadow grass; *Poa nemoralis*, woodside meadow grass; *Trifolium repens perenne*, perennial white clover; *Trifolium minus*, yellow suckling. Sow the mixture at the rate of 3 bushels (60 lb.) to the English acre, or 1 gallon (2½ lb.) to 6 rods or perches.

For a STIFF SOIL RESTING ON CLAY, a good mixture would consist of *Poa pratensis*, smooth-stalked meadow grass; *Poa trivialis*, rough-stalked meadow grass; *Lolium perenne tenue*, fine rye grass; *Festuca duriuscula*, hard fescue; *Trifolium repens*, white clover; *Trifolium minus*, yellow suckling.

For a LIGHT SANDY SOIL the mixture should consist of, or at least comprise, *Lolium perenne tenue*, *Poa pratensis*, *Festuca duriuscula*, *Avena flavescens*, the yellowish oat grass; *Trifolium repens*, *Lotus corniculatus*, the bird's-foot; *Achillea millefolia*, the common yarrow.

For a THIN SOIL RESTING ON CHALK OR LIMESTONE, the selection should comprise *Lolium perenne tenue*, *Festuca duriuscula*, *F. ovina*, *Poa trivialis*, *Cynosurus cristatus*, *Medicago lupulina*, yellow medick; *Trifolium repens*, *T. minus*, *Lotus corniculatus*.

For a collective prescription we cannot do better than adopt that recommended by Messrs. Lawson and Son, the eminent seedsmen of London and Edinburgh. The several quantities of the several sorts named constitute a mixture for one English acre.

	LIGHT SOIL.	MEDIUM SOIL.	HEAVY SOIL.
Avena flavescens	1 lb.	0 lb.	0 lb.
Cynosurus cristatus	5	6	7
Festuca duriuscula	3	3	4
Festuca tenuifolia	2	2	1
Lolium perenne tenue	20	20	20
Poa nemoralis	1½	1¾	2
Poa nemoralis sempervirens	1½	1¾	2
Poa trivialis	1½	1¾	2
Trifolium repens	7	7	7
Trifolium minus	2	2	1

Under trees a little variation of the mixture must be adopted. Leave out the two species of Fescue, and substitute similar quantities of *Poa nemoralis*. Indeed, *P. nemoralis angustifolium* is the best of all grasses to produce a beautiful sward under trees, its growth being so close that it displaces weeds, and it is green in spring earlier than most other grasses; and as it also does well in exposed places, it may be made "a note of," for any one in a state of distress at the shabbiness of a lawn. Another most useful lawn grass is *Lolium perenne tenue;* but, as it is the twin brother of that very worst of lawn grasses, *Lolium perenne,* or common rye grass, care must be taken to obtain the right sort. It thrives on almost any soil that is not wet, and is delightfully fresh all the winter.

TREE VIOLET.

CHAPTER XVII.

GARDEN VERMIN.

THE best general advice that can be given on the subject of garden vermin would be to this effect—grow your plants well, and you will be very little troubled with vermin. We need not discuss the philosophy of the matter, but matter of fact it is, that healthy vigorous vegetation is rarely assailed by destroying insects, and on the other hand, vegetation in a diseased or starving state will be attacked by many plagues, including moulds and mildews, in addition to aphis, thrip, scale, and red spider. We may safely say that amateurs frequently invite the small marauders by their mismanagement, but it would be unjust to say that the outbreak of a plague in the garden is invariably an evidence of the cultivator's neglect or error. It is, however, of the utmost importance for the amateur to bear in mind that in any and every case an insect enemy is to be met in the first instance by any means that will promote the vigour of the plants attacked. In order the more clearly to be understood, let us suppose that we are requested to advise on a plantation of roses infested with green-fly, mildew, and thrip. We will further suppose that the weather has been hot and dry for some time, and that the leaves of the rose-trees are yellowish, and the growth of the season is pushing in a weak and wiry manner, instead of rising in strong shoots, indicative of a vigorous root action, what should we advise in such a case? Just such a course of procedure as follows:—First soak the roots well every alternate evening for a week, each separate soaking to be an imitation of the deluge. On the evenings when there is no water given to the roots, give them a heavy shower from a powerful engine, taking care to send the water with some force through the heads of the trees, so as to drench the under as well as the upper sides of the foliage.

At the end of the week give one more heavy soaking, and the next day spread over the ground amongst the trees a coating of rotten manure; or a mixture of fine earth and guano; or a mixture of guano, wood-ashes, and earth. This course of treatment would annihilate the vermin, and put new vigour into the trees, and establish a most valuable rule for future action, founded on this fact, that *pure soft water is the most potent of all insecticides.*

Let us suppose another case. Say a lot of asters, stocks, or solanums in a pit or frame, smothered with green-fly. If we had to advise upon them, we should first consider the state of the weather. If still cold and unsettled, we would syringe the plants with a weak solution of "Fowler's Insecticide," and give the roots a soaking of weak manure water. But if we could trust the weather, we should advise the planting of them out in the open ground, with a good watering to follow to start them into growth, and escape the trouble and expense of employing a preparation. They would "grow out" of fly, as we say in garden phraseology, and almost by magic become clean and thrifty.

But it will frequently happen that vermin will attack plants that appear to be in perfect health, and in such a case the resort to invigorating measures may appear to be a misdirection of our energies. At the risk of appearing tedious, we will here remark that plants, apparently full of vigour, may be suddenly debilitated by excessive heat or drought, and in such a case invigorating measures may be really needed, although there are no striking evidences in the appearance of the plant of an impoverishment of its juices. Therefore we propose that the cultivator should be cautious against delusions, and as much as possible make it a rule to invigorate the plant that is beset with vermin, as the first step towards a purification.

However, one way or another, we come to nostrums at last, and the first amongst many that we shall recommend for general adoption, is—

HOT WATER, which we have proved, by numerous experiments, to be more efficacious than cold water as a vermin killer, and a perfectly safe insecticide if employed in a sensible manner. The "monarch of all I survey" in the world of plant-vermin is the Aphis. Now, hot water will unseat this king and lay him in the dust. All vermin love dirt and hate

water; but this potentate particularly objects to a warm bath
as poison to him. Beware, however, that in scalding him to
death you do not parboil your plants. Any plant in a growing
state may be dipped into water heated to 120° Fahr. without
the slightest harm, and if kept in a state of immersion a few
seconds, every aphis upon it will surely perish. In a series of
experiments carefully conducted in our garden at Stoke Newington, we found that fuchsias in a growing state were unhurt
if immersed in water heated to 140° Fahr.; that at the same
temperature, calceolarias and Chinese primulas were seriously
injured; that pelargoniums of all kinds were unhurt if
plunged for several minutes in water heated to 150°; that
centaureas, sedums, saxifragas, heliotropes, petunias, begonias, mignonette, and many other plants of soft texture,
could endure a temperature of 140° without the slightest
harm; but at 150° they suffered more or less. About ten
years ago we reported in the "Floral World" that Fairy or
Lawrence roses, which are grown in quantities in pots for
market, could be best kept clean by dipping in hot water, as
at 120° the plant is not injured, and every aphis upon it is
destroyed. This simple method of removing vermin from
plants is, we are quite satisfied, capable of very general
adoption, in place of more troublesome and more expensive
plans.

TOBACCO POWDER, as prepared by the sanction of the
Excise, we place next in order of value, for it is cheap, convenient, and cleanly in use, harmless to vegetation, but most
deadly to aphis and thrips. There appears to be no important difference of value between "Pooley's," "The London,"
"Fowler's," and other preparations; but in any case tobacco
powder should emit an unmistakable odour of sulphur, for
while the Excise require it to be spoilt for use as snuff, the
addition of sulphur greatly increases its efficiency as a
vermin destroyer. Tobacco powder should be dusted on
the leaves of the infested plant when they are damp with
dew, and should be washed off twelve hours afterwards.

The best of the many liquid preparations offered to a
"discerning public," are the APHIS WASH, manufactured by
the "City Soap Company," and FOWLER'S INSECTICIDE. It
is sufficient to say of them that they must be employed in
strict accordance with the printed directions that accompany
them, and that they are equally safe and effectual.

One of the most destructive insects the gardener has to wage war against is the "Daddy Longlegs" (*Tipula oleracea*), the grub of which may be likened to a minute sausage of a black colour, with almost invisible feet. This "leather-jacket" wretch feeds at the surface of the soil; usually above ground, never below. The result of this marauding is that the plants he feeds upon are nipped through "between wind and water," as a sailor would say, and they perish very soon afterwards, very much sometimes to the bewilderment of the inexperienced gardener. This is a most difficult pest to deal with, and unfortunately it occurs in profuse abundance in some seasons, and is so very partial to grass turf, that it will pretty well destroy a lawn in the course of a week or two if allowed. When garden plants are seen to fall over mysteriously, the cultivator should carefully stir the earth about their roots, to hunt up the dark leather-jacket buried in the soil, an inch or so deep. If he is found to be the cause of the death of the plant, the best course to pursue will be to hoe the soil carefully all over the bed, and finally to remove the soil with a piece of stick, so as to form a shallow basin around every plant. This simple course of procedure is remarkably effective in reducing the numbers of the daddy-longlegs grubs, and we are quite unable to explain the reason why; though we have always supposed that the small birds, the robins perhaps in particular, having a propensity to examine newly-disturbed soil, make a rush to the bed that has been operated upon in this way, and find the leather-jackets before they have time to hide themselves in new burrows. When these marauders take possession of a lawn and threaten to destroy it, they can be dealt with in a most direct and destructive manner. Every night the lawn should be rolled with a heavy roller, and every morning at earliest dawn the lawn should be sprinkled with quicklime. The roller will crush myriads that have come out to feed, and the lime will kill myriads that have escaped the roller.

It is by no means an unwise plan to feed some kinds of vermin, for that plan, when it can be adopted conveniently, makes an end of a vast amount of bother. For example, we are about to plant out a bed of dahlias, and we know that snails and slugs will pounce upon them the very first night they are planted, and probably make an end of them for ever. To prevent that, we plant the whole piece with lettuces first,

18

and so long as the lettuce lasts, the snails and slugs will not touch a dahlia. There is much to be done in this way without incurring the risk of multiplying the vermin on a place, because your provisions may be turned to account as traps, and by examining the lettuces every night by the aid of a lamp, you may bag all the snails in the district, and at the end of the undertaking have a lot of nice lettuce for the table. For a last word, we say, encourage the small birds, for they are wonderful aids to the amateur gardener as destroyers of insects. The sparrows may perhaps vex you by nibbling your crocuses, and the blackbirds may steal your cherries; but remember they cannot trouble you in this sort of way all the year round. They will be gobbling up snails and caterpillars and butterflies in the dewy dawn, when you perhaps are sleeping and unhappily unconscious of the benefactions of your feathered friends.

To catch and kill vermin must be the constant duty of every amateur gardener. The large marauders, such as snails and woodlice, will never cease from troubling, and it may be matter for thankfulness that they pronounce a dread sentence against the gardener who, in the midst of work, goes to sleep. You must catch the vagabonds. Go to work in this way. Lay little heaps of lettuce leaves in cool, quiet places, and examine them at dusk and daybreak. Catch and kill in any way you please; a pot of brine is a very good bath for the purpose. Lay about also in the neighbourhood of choice subjects to which snails are partial, nice young cabbage leaves slightly smeared with rank butter; catch and kill as before. Lay about in cool quiet spots small heaps of fresh brewers' grains; catch and kill as before. Where woodlice abound, take some *dirty* flower-pots (always combine a little dirt of some sort with vermin traps, for vermin are extravagantly fond of dirt), and fill these dirty flower-pots with dry moss and crocks mixed together. Place them where the vermin abound, and cover each with a tuft of dry moss. Every other day, proceed to catch and kill in this simple manner. Have a large pailful of boiling water. Take up a pot quietly, and quickly shoot out its contents into the water. You clear away your enemies by thousands in this way; there is no trap to equal a dirty pot filled with dirty crocks, and dirty (but dry) moss. If there arises any peculiar difficulty, such as a choice plant being eaten nightly, and you cannot

catch the marauder, take a slice of apple, and surround it
with dry moss, in a flower-pot.   Take also a slice of potato,
and use it in the same way.   Place these two pots one
on each side of your delicate subject, but at a distance
of six inches or so, and at dusk and dawn turn out the
contents of each pot quickly, and it will be a strange thing if
the marauder is not bagged.

GARDEN TOOL-HOUSE AND READING-ROOM.

CHAPTER XVIII.

IN the several chapters and their proper sections are included
lists of the most distinct and useful species and varieties of
garden plants. But the lists that follow may prove acceptable as supplementary to the foregoing, as adapted to particular cases.

A FEW HARDY PLANTS FOR WALLS AND BOARDED FENCES.
—*Pyracantha*, most beautiful in autumn, with its fiery berries.
Virginian Creeper; the small-leaved sorts are the best. *Aris-
tolochia*, fine large leaves; requires a warm aspect. *Dutch
Honeysuckle* and *Scarlet Trumpet Honeysuckle*. *Clematis
flammula*, best adapted for arbours and gateways. *Clematis
rubro-violacea*, splendid flowers. *Pyrus japonica*, will flower
freely on a north wall. *Ivies*, in variety, the most handsome
being common *Irish*, common *English*, *Algerian*, and the
Thick-leaved; the *variegated-leaved ivies* are sumptuous wall
plants, well adapted for north and west aspects.

CHEAP AND HARDY PLANTS FOR SHADY BORDERS.—All the
double and single *Primulas*, *Polyanthuses*, *Auriculas*, and
Pansies; Solomon's Seal; Periwinkles in variety; *Anemone
vitifolia; Columbines* in variety; *Aspidistra lurida; Ruscus
aculeatus; Campanula carpatica* and *pumila; Lily of the
Valley; Funkia Sieboldiana*, and half-a-dozen others. *Christmas Rose; Day Lily; Yellow Moneywort* and *Lysimachia
thyrsiflora; Forget-me-not; Double Narciss* and *Hoop-Petti-
coat Narciss; Œnothera riparia; Polemonium cœruleum;
Saxifraga hypnoides, oppositifolia*, and several others; *Scilla
campanulata; Symphitum coccineum; Spider-wort; Spiræa
japonica*.

A SELECTION OF TREES AND SHRUBS FOR ENTRANCE
COURTS. — Deciduous Trees of moderate growth: *White-
leaved Negundo (Acer fraxinifolium*, var.); *Cut-leaved Alder;*

Snowy Mespilus; Common Almond; Double-flowering Peach; Berberis asiatica; Common Birch; Bird Cherry; Variegated Dogwood (*Cornus mascula variegata*); *Cotoneaster Simmondsi; Common Laburnum; Persian* and *Common Lilac; Deutzia scabra; Euonymus europœus* (will thrive in the deepest shade of large trees); *Common* and *Venetian Sumach.* Evergreen Trees and Shrubs : *Aucuba japonica; Berberis Darwini; Common Tree Box* and *Round-leaved Box; Euonymus japonica,* and its golden-leaved varieties (a splendid series); *Green* and *Variegated Hollies; Chinese Privet; Phillyrea; Holly-leaved Evergreen Oak; Skimmia japonica; Laurestinus.*

A FEW GOOD HARDY FLOWERING PLANTS FOR A SMALL GARDEN.—All the sorts named in this list require sunshine, more or less, but in situations where there is a free circulation of air, the partial shade of a few neighbouring trees will scarcely affect them. *Achillea filipendula,* and *A. ptarmica; Agrostemma coronaria; Alyssum saxatile; Anemone japonica,* and *A. stellata; Antirrhinums* in variety, for dry sunny spots ; *Columbines* in variety; *Common White Arabis; Armeria dianthioides; Arundo conspicua; Aster elegans,* and *A. amellus; Purple Aubrietia; Callirhoe involucrata; Campanula carpatica, C. glomerata, C. persicifolia,* and *C. pyramidalis; Corydalis aurea; Delphinium formosum; D. Hendersoni,* and others. *Dianthus plumarius, D. hybridus, Common Clove,* and *Pinks* in variety. *Eryngium amethystinum; Fritillaria imperialis; Helianthus multiflorus; Hepaticas* in variety. *Iberis sempervirens; German Iris* in variety. *White Everlasting Pea* makes a grand covering for a mound. *Hybrid Clematis,* in variety make splendid beds, requiring only once planting for a life-time. *Lilies* in variety, the most desirable being *Common White, Longiflora, Scarlet Martagon, Tigrinum, Excelsum,* and *Bulbiferum; Lychnis dioica rubra* and *L. viscaria plena. Œnothera fruticosa* and *Œ. macrocarpa; Papaver bracteatum; Hybrid Phlox,* in variety; *Hybrid Potentilla,* in variety; *Primula acaulis* and *elatior,* in variety; *Hybrid Pyrethrums,* in variety; *Hollyhocks,* in variety; *Spiræa filipendula, S. venusta,* and *S. palmata; Tritoma uvaria; Trollius asiaticus.*

HARDY PLANTS FOR WATER SCENES.—*Butomus umbellatus,* flowering Rush; *Caltha palustris flore pleno,* double marsh Marigold; *Calla palustris,* marsh Trumpet Lily ; *Iris pseudo-acorus,* yellow Water-flag; *I. Sibirica,* Siberian Iris; *Sagittaria sagittifolia,* common Arrowhead; *S. cordifolia,* heart-

leaved Arrowhead ; *Menyanthes trifoliata*, three-leaved Buck-
bean ; *Lythrum salicaria*, willow-leaved Loosestrife ; *Narthecium
ossifragum*, Lancashire Asphodel ; *Potamogeton natans*, floating
Pond-weed ; *Villarsia nymphæoides*, lily-like Villarsia ; *Acorus
calamus*, sweet Flag ; *Arundo donax*, Reed-grass ; *Carex
maxima*, greater Sedge ; *Scirpus radicans*, Club-rush ; *Comarum
palustre*, marsh Cinquefoil ; *Cyperus longus*, tall Sedge ; *C.
virens*, green Sedge ; *Juncus conglomeratus variegatus*, variegated Rush ; *J. effusus spiralis*, spiral soft Rush ; *Lysimachia
thyrsiflora*, thyrse-flowered Lysimachia ; *Nasturtium sylvestris*,
woodland Water-cress ; *Nuphar lutea*, yellow Water Lily ;
Nymphæa alba, white Water Lily ; *Pilularia globulifera*,
globular Pillwort ; *Poa aquatica*, Water-grass ; *Sparganium
ramosum*, branching Bur Reed ; *S. minimum*, lesser Bur Reed ;
Stratiotes aloides, Water Soldier ; *Typha latifolia*, broad-leaved
Reed-mace.

HARDY HERBACEOUS PLANTS FOR PLANTING ON ROCKERIES.
—*Ajuga reptans variegata*, creeping Bugle ; *Linaria cymba-
laria*, Ivy-leaved Toadflax ; *Dianthus petræus*, Rock Pink ;
Diotis maritima, seaside Diotis ; *Doudia (Hacquetia) epipactis*,
yellow Doudia ; *Geranium sanguineum*, blood-flowered Geranium ; *G. Lambertianum*, Lambert's Geranium ; *G. striatum*,
striated Geranium ; *Nepeta violacea*, Violet Mint ; *Ononis
rotundifolia*, round-leaved Rest-harrow ; *Aubrietia erubescens*,
blush-flowered Aubrietia ; *Draba aizoides*, Whitlow Grass ;
D. borealis, Northern Whitlow Grass ; *Iberis corrœafolia*,
Corræa-leaved Candytuft ; *Acæna Zealandica*, New-Zealand
Acæna ; *Potentilla nemoralis*, Wood Cinquefoil ; *P. alpestris*,
alpine Cinquefoil ; *Statice armeria alpina*, alpine Thrift ; *S.
bellidifolia*, daisy-leaved Thrift ; *S. eximia*, choice Thrift ;
Lotus corniculatus minimus, small Bird's-foot Trefoil ; *L. c.
flore pleno*, double-flowered small Bird's-foot Trefoil ; *Lychnis
dioica*, diæcious Lychnis ; *L. Chalcedonica*, Chalcedonian
Lychnis ; *L. flos Jovis*, Jove's Lychnis ; *Agrostemma githago*,
Corn Cockle ; *A. coronaria*, Rose Campion ; *Antirrhinum
majus* (in var.) Antirrhinum ; *A. sempervirens*, everlasting
Antirrhinum ; *Soldanella alpina*, alpine Shilling Leaf.

HARDY HERBACEOUS PLANTS FOR DRY BARREN PLACES.—
Aspidistra lurida variegata, lurid Shield Lily, handsome leaves ;
Sternbergia lutea, autumn Crocus ; *Othonna cheirifolia*, wall-
flower-leaved Ragwort ; *Eryngium amethystinum*, amethyst
Sea Holly ; *Oxalis corniculata rubra*, pure-leaved Horned

Oxalis; *Trifolium repens* var. *nigrescens*, black-leaved Clover; *Silene alpestris*, alpine Catchfly; *Megasea rubra*, broad-leaved Saxifrage; *M. ciliata*, hairy, broad-leaved Saxifrage; *M. cordifolia*, heart-shaped-leaved Saxifrage; *Aster bicolor*, two-coloured Starwort; *Linaria alpina*, alpine Toadflax; *Antennaria alpina*, alpine Everlasting; *A. dioica*, diæcious Everlasting; *Ferula asperifolia*, Giant Fennel; *Eryngium Bourgati*, Sea Holly; *Yucca recurva*, recurved Adam's Needle; *Y. gloriosa*, glorious Adam's Needle; *Y. filamentosa*, thready Adam's Needle; *Iris reticulata*, reticulated Iris; *Santolina squarrosa*, lavender Cotton; *S. chamæcyparissus*, ground Cypress; *S. viridis*, dark green lavender Cotton; *Helianthemum venustum*, comely Sunrose; *H. serphyllum* (in var.), thyme-leaved Sunrose; *H. rosmarinifolium*, rosemary-leaved Sunrose; *H. tuburarium*, Italian rosemary-leaved Sunrose; *H. globulariæfolium*, globular-leaved Sunrose; *Elymus arenarius*, Lyme-grass; *E. mexicanus*, Mexican Lyme-grass; *Arabis lucida*, shining Arabis.

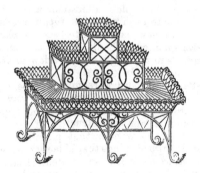

CHAPTER XIX.

JANUARY.—Earth-work and improvements must be regulated by the weather, but at every opportunity must be pushed on with all possible speed, for there will soon be other work to do. Stack up turf for composts. Spread a mulch of half-rotten manure amongst newly-planted shrubs and roses. Renovate flower-beds by deep digging and manuring. Prepare beds for ranunculuses and anemones. Any autumn-planted bulbs that are pushing through should be covered with a sufficiency of earth to protect them for awhile from severe frost.

FEBRUARY. — Finish planting deciduous trees as soon as weather permits, and complete alterations and improvements. This is a good time to form rockeries, repair roads, put down edgings, and make new lawns. Herbaceous plants may be divided and transplanted. Plant ranunculuses and anemones. Sow hardy annuals in the borders, and a few in frames to be transplanted for an early bloom.

MARCH.—Beds and borders requiring to be cleaned must be carefully dealt with, to spare from injury any plants that are pushing through. The routine plan of digging mixed borders in spring is simply destructive of the good plants they contain. It is no wonder that pæonies, dielytras, delphiniums, and other good things that hide themselves in winter, invariably vanish from borders so treated. Plant herbaceous plants. Sow seeds of annuals and perennials. Look over rose-beds to make stakes safe, and tread firm any that are loose at the roots. Carefully hoe the top crust amongst tulips, pansies, and other choice subjects that are planted in beds. When the pansies have been cleaned, peg down the main branches, and strew over the bed a mixture of fine earth and rotten manure to promote surface roots.

APRIL.—Prune ivy to one layer of branches, and remove

all the leaves. The new growth will follow instantly, and be even and rich; continue to sow seeds of hardy subjects in open borders, and of tender subjects in frames. Part chrysanthemums, and strike cuttings of sorts that are required in quantity. Plant dahlia roots. Plant Tigridia bulbs. When tulips are frozen hard, water them with cold water before the sun shines on them.

MAY.—Sow in open borders asters, balsams, and other half-hardy subjects and hardy annuals to succeed the early sowings. Plant dwarf roses out of pots; this being one of the best methods of forming a plantation, secure them on their own roots if possible. Thin and stake flowering shoots of carnations and pinks. Bedding out should not commence until towards the end of the month, unless the plants are known to be well hardened.

JUNE.—Bedding out is the principal business now. Take the plants in the order of their relative hardiness, so as to keep back to the last moment all the more tender subjects, such as coleus, alternanthera, etc. Keep lawns in the best order possible, and in the event of prolonged dry weather, flood them with water at least once a week. If the grass is thin and poor, remove the box from the mowing machine, that the mowings may be scattered. Strike pipings of carnations and pinks. Strike cuttings of pansies from the young shoots. Stake and mulch dahlias. Look over mixed borders, and provide supports for plants that are likely to be blown over.

JULY.—Watering is an important business now. As a rule geraniums, centaureas, antirrhinums, lantanas, salvias, sedums, and sempervivums planted out thrive better without artificial watering than with it, though when first planted, one or two good soakings may be needed to give them a start. On the other hand, calceolarias, cannas, coleus, dahlias, hollyhocks, carnations, lobelias, heliotropes, and verbenas, will be benefited by copious watering in dry weather, but mere surface driblets will rather injure than advance their welfare. Bud roses. Layer carnations. Take cuttings of bedding plants as fast as they can be obtained of a proper size and substance, and strike them without aid of artificial heat. Put earwig traps on dahlia stakes.

AUGUST.—Propagate bedding plants in quantity. Plant out carnations and pinks that were struck early from pipings. Sow seeds of hardy perennials. This is a good time to sow grass seeds for the formation of new lawns, but if turf is to laid, wait until next month.

SEPTEMBER.—Lay down grass turf. Plant box-edgings, and all kinds of evergreen shrubs. Prepare for planting bulbs. All kinds of lilies may now be transplanted. This is the best time for striking cuttings of calceolarias. Plant hardy herbaceous plants.

OCTOBER.—Plant hardy bulbs and tubers of all kinds. Look over chrysanthemums to insure their timely and sufficient supports. Take up pompones required for " plunge beds," and pot them, taking care to injure the roots as little as possible.

NOVEMBER. — Keep chrysanthemums in good order, and securely staked. Remove into pits and frames all the nearly hardy plants that require shelter, not so much because of danger from frost as yet, but to protect them from heavy rains. Provide protection for plants of questionable hardiness that cannot be removed to frames. Take up dahlia roots as soon as the tops are killed by frost. Plant standard roses and briers for budding next season. Continue to plant bulbs. Plant deciduous trees of all kinds.

DECEMBER.—Finish planting bulbs. Hyacinths, tulips, and narcissi planted now will flower well in April next. There must, however, be no delay, or the season will be lost. Cut down hardy fuchsias. Spread a thin coat of dry flaky manure over beds of choice pansies, carnations, pinks, tulips, pentstemons and phloxes, both to protect from frost, and afford nourishment by the solvent action of snow on the manure.

SPERGULA TURF (ARENARIA CÆSPITOSA).

INDEX.

Printed in the United States
By Bookmasters